AF667268

INVERSE AND ILL-POSED PROBLEMS SERIES

Multidimensional Inverse and Ill-Posed Problems for Differential Equations

Previously published in the
Inverse and Ill-Posed Problems Series:

Inverse Problems for Maxwell's Equations
V.G. Romanov and S.I. Kabanikhin

Integral Geometry of Tensor Fields
V.A. Sharafutdinov

Ill-Posed Problems with A Priori Information
V.V. Vasin and A.L. Ageev

Other related books from VSP:

Inverse Problems of Mathematical Physics
V.G. Romanov

Inverse Sturm-Liouville Problems
B.M. Levitan

Related Journal:

Journal of Inverse and Ill-Posed Problems
Editor-in-Chief: M.M. Lavrent'ev

INVERSE AND ILL-POSED PROBLEMS SERIES

Multidimensional Inverse and Ill-Posed Problems for Differential Equations

Yu.E. Anikonov

///VSP///

Utrecht, The Netherlands, 1995

VSP BV
P.O. Box 346
3700 AH Zeist
The Netherlands

© VSP BV 1995

First published in 1995

ISBN 90-6764-185-5

All rights reserved. No part of this publication may be reproduced, stored in a retrieval system, or transmitted in any form or by any means, electronic, mechanical, photocopying, recording or otherwise, without the prior permission of the copyright owner.

CIP-DATA KONINKLIJKE BIBLIOTHEEK, DEN HAAG

Anikonov, Yu.E.

Multidimensional inverse and ill-posed problems for
differential equations / Yu.E. Anikonov. - Utrecht : VSP
- (Inverse and ill-posed problems series, ISSN 1381-4524)
With ref.
ISBN 90-6764-185-5 bound
Subject headings: inverse problems / ill-posed problems
/ differential equations

Printed in The Netherlands by Koninklijke Wöhrmann bv, Zutphen.

CONTENTS

INTRODUCTION

The problem of determining differential equations by information on solutions of these equations is called the inverse problem for differential equations.

Examples of these kind of problems are the well-known Sturm-Liouville problems. Many important application problems connected to elastic displacement, electromagnetic oscillation, diffuse and other processes in nature and society lead to inverse problems. The extent of these problems is expanding constantly. At present a substantial amount of literature is devoted to these problems. One of the first investigations into these kind of questions was the inversion of kinematic problems of seismics, the essence of which consists in the determination of the velocity of propagation of elastic waves by time of their movement. A one-dimensional case of one of such problems was considered by Herglots (1905). He obtained the formula of inversion, which later became the basis of the solution of many important problems in geophysics, in particular the basis of structure determination of the Earth's crust and the Earth's mantle.

As a second classic direction in the theory of inverse problems one can mention the inverse problem of the theory of potential, which consists in a form of description of body shape and density of this body on a known potential. The uniqueness theorem of the solution of one of these problems was proven by Novikov (1938) for the first time.

Problems linked to the Sturm-Liouville equation and its generalization are a third direction in the theory of inverse problems. The sense of these problems is the following: we know the spectral function or scattering data of a differential operator; it is required to define this operator. The first uniqueness theorems of the solution were obtained in papers by Ambarzumjan (1929) and Borg (1945).

Inverse problems are usually nonlinear and are separated into one-dimensional and multidimensional problems, depending on whether the sought function (or functions) is a function of one variable or of many. These problems, especially multidimensional problems, are often ill-posed in the classic sense. In this sense, questions of the uniqueness of the solution and a search of minimal information, which makes the inverse problem determined, have particular actuality. The theorem of uniqueness of the solution of the complex multidimensional inverse problem for the Schrödinger equation in the class of piecewise-analytical functions was established for the first time by Berezanskij (1958).

Multidimensionality of inverse problems has particluar value at present, because practice shows that many investigating processes are described by an equation, of which the coefficient essentially depends on many variables. Inverse problems often bring integrals to first kind operator equations. Some inverse problems for hyperbolic equations are, for example, reduced to the investigation of integral equations of Volterra type of the first kind. In turn, this sometimes allows one (chiefly in one-dimensional inverse problems) to find equations of the second kind with operators possessing sufficiently good properties, for example compact operators, which gives a method of investigation. The basis of such a convergence is often a formula for the solution of direct or inverse problems. In many cases, particularly when information on the solution of an equation is given by only a part of the boundary of the considered domain (practice is needed in those problems), such a convergence of the inverse problem to an integral equation of the second kind often appears to be impossible. One of the reasons for this is the ill-posedness of these problems. These questions require new approaches. The general theory of operator equations of the first kind and their applications was developed in papers by Tikhonov (1943, 1963), Tikhonov and Arsenin (1977), Lavrent'ev (1955, 1959) and Ivanov (1963). The new methods developed there, have found wide implementations, in particular in the theory and practice of ill-posed problems. Many inverse problems are closely connected to problems of integral geometry. Thereby, it appears necessary to investigate new problems of integral geometry, when manifolds in which the sought function (or functions) is integrated, are complicated by their structure. Important results and applications in the case of linear and other manifolds are obtained from papers by Radon (1917), Courant and Hilbert (1962), John (1955), Khachaturov (1954), Kostelyanec and Reshetnyak (1954), Gel'fand.(1960), Gel'fand et al. (1966), Helgason (1959), Semyanistyi (1966) and others. At present these results on integral geometry and other applications have found applications in inverse problems, in particular in tomography. Multidimensional problems of integral geometry in the case of complex manifolds and their connection to inverse problems are formulated and investigated in articles by Lavrent'ev et al. (1970, 1986), Romanov (1987), Anikonov (1987b), Anikonov and Pestov (1990a), and Bukhgeim (1983, 1986, 1988). A significant contribution to the theory of inverse problems has also been made by Prilepko and Kostin (1993), Isakov (1990), Blagoveshchenskii (1986), Anikonov (1975), Anikonov et al. (1993), Lorenzi (1992), Yamamoto (1990, 1992), Anger (1990), Sabatier (1990a,b) and others.

This monograph is devoted to statements of multidimensional inverse problems, in particular to methods of their investigation. Questions of the uniqueness of solution, solvability and stability are studied. Methods to construct a solution are given, and in certain cases inversion formulas are given as well. Concrete applications of the theory developed here are also given. Where possible, we have stopped to consider the method of investigation of the problems, thereby sometimes losing generality and quantity of the problems, which can be examined by such a method.

This work was supported by the Russian Foundation for Fundamental Research, grant No. 93-011-181.

CHAPTER 1

Operator Equations and Inverse Problems

1.1 DEFINITION OF QUASIMONOTONICITY, THE UNIQUENESS THEOREM

The investigation of inverse problems for differential equations frequently reduces to the investigation of the first type operator equations. Such operators may come up as functionals of the solutions of some differential equation correspond to this equation. The other words, the operator is defined as the set of equations and its values are functionals of the solutions of these equations. It is often found that the equation is defined simply by one real function belonging to some class. So the operator of the inverse problem is defined at the set of the functions. The operators of inverse problems may be of complex nature; they are nonlinear as a rule. Some of them have the property that if one of the functions is larger than another in a subset of the domain of definition of these functions, then images of these functions are different.

It was this property which was used for the proof of the solution uniqueness of one-dimensional inverse problems of electroprospect in the paper of (Tikhonov, 1949). In the multidimensional case (Berezanskii, 1958) applied this property for the proof of the solution uniqueness of the inverse problem of the Schrödinger equation in the class of the piecewise analytical functions. We pick out the operators by such a definition.

Let E be a set of elements x, and let $\{\lambda\}_E$ be some set of real functions $\lambda(x)$ in E. Suppose that an element τ belonging to some set $\{\tau\}$ corresponds to every function $\lambda(x) \in \{\lambda\}_E$ by $\tau = \mathbf{M}\lambda$, where $\mathbf{M}$ is some operator.

As a covering of set E we understand totality $\{\omega\}$ of sets $\omega \in E$, union of which is equal to E.

Definition. Operator $\mathbf{M}$ is called quasimonotonic with respect to the covering $\{\omega\}$ of set E, if it follows from inequality $\lambda_1(x) > \lambda_2(x)$ holding for all x belonging to, at least, one non-empty set $\omega \in \{\omega\}$ that $\mathbf{M}\lambda_1 \neq \mathbf{M}\lambda_2$, $\lambda_i(x) \in \{\lambda\}_E$, i=1,2.

Lemma 1.1. *If for every $\tau_0 \in \{\tau\}$ the equation $\mathbf{M}\lambda = \tau_0$ has a unique solution, then the operator $\mathbf{M}$ is quasimonotonic with respect to any covering $\{\omega\}$ of set E.*

This statement is obvious.

Our main purpose is to show that in special cases it follows from being quasimonotonic with respect to a fixed covering that the solution of the equation $\mathbf{M}\lambda = \tau_0$ is unique.

Let $\mathbf{R}^{n+1}$ be an $(n+1)-$ dimensional real Euclidean space for (x,y), $x \in \mathbf{R}^n$, $-\infty < y < \infty$, $n \geq 0$, let E and $\bar{E}$ be semispaces with $y > 0$ and $y \geq 0$. Denote by $\{\lambda\}$ the set of all real infinite differentiable functions $\lambda(x,y)$, $(x,y) \in \bar{E}$ to be quasianalytical in y in this domain. Under the covering $\{\omega\}$ of the semispace E we understand the set of all hemiballs $\omega(\xi,p)$:

$$\omega(\xi,p) = \{(x,y) : |x-\xi|^2 + y^2 < p^2, \quad y > 0\}, \quad \xi \in \mathbf{R}^n, \quad p > 0.$$

Lemma 1.2. *For every function $\lambda(x,y) \in \{\lambda\}$, which is not identically equal to zero, there exists a hemiball $\omega \in \{\omega\}$ and constants $c > 0$, $\alpha \geq 0$ such that the inequality*

$$|\lambda(x,y)| \geq cy^\alpha, \quad (x,y) \in \{\omega\}$$

holds.

Proof. Since the function $\lambda(x,y)$ is not identically equal to zero, and quasianalytical in y in $\bar{E}$, the equality

$$\left.\frac{\partial^k \lambda}{\partial y^k}\right|_{y=0} = 0$$

is impossible for all k, $k = 0,1,2,\ldots$. So there exists a smallest number $m \geq 0$ and a point $\xi \in \mathbf{R}^n$ such that

$$\frac{\partial^k \lambda(x,0)}{\partial y^k} = 0, \quad k = 0,1,2,\ldots,m-1, \quad x \in \mathbf{R}^n, \quad \frac{\partial^m \lambda(\xi,0)}{\partial y^m} \neq 0. \tag{1.1}$$

If $m = 0$,then the statement is obvious.

Let $m > 0$. Every infinite differentiable function $\tilde{\lambda}(x,y)$ such that $\tilde{\lambda}(x,0) = 0$ (see (Malgrange, 1966)) allows a representation $\tilde{\lambda}(x,y) = y\tilde{h}(x,y)$, where

$$\tilde{h}(x,y) = \int_0^1 \frac{\partial \tilde{\lambda}(x,ty)}{\partial y} dt. \tag{1.2}$$

Combining (1.1) and (1.2) we have

$$\lambda(x,y) = y^{m-1} h(x,y), \tag{1.3}$$

where $h(x,y)$ is an infinite differentiable function and $h(\xi,0) \neq 0$. Let $\overline{R}(\xi,p)$ be a closed ball in $\mathbf{R}^{n+1}$ of radius $\rho > 0$, centered at $(\xi,0)$ such that $h(x,y) \neq 0$, $(x,y) \in \overline{R}(\xi,p)$. By continuity $h(x,y)$ and $h(\xi,0) \neq 0$ such ball exists.

Let $c = \min_{(x,y)\in\overline{R}(\xi,p)} |h(x,y)|$, and $\omega = R(\xi,p) \cap E$. By (1.3) we have

$$|\lambda(x,y)| = y^{m-1}\,|h(x,y)| \geq cy^{m-1}, \quad (x,y) \in \omega.$$

The Lemma is proved.

By this Lemma it follows

Theorem 1.1. *Let $\mathbf{M}$ be a quasimonotonic operator with respect to the covering $\{\omega\}$ of semispace E by hemiballs $\omega(\xi,p)$. If $\mathbf{M}\lambda_1 = \mathbf{M}\lambda_2$, $\lambda_i(x,y) \in \{\lambda\} \subset \{\lambda\}_E$, then $\lambda_1(x,y) = \lambda_2(x,y)$ in E.*

PROOF. If $\lambda_1(x,y) \neq \lambda_2(x,y)$, then by Lemma 1.2 there exists a hemiball $\omega \in \{\omega\}$ such that $|\lambda_1(x,y) - \lambda_2(x,y)| \geq cy^\alpha > 0, \quad (x,y) \in \omega$. In particular it follows from this that in the domain ω either $\lambda_1 > \lambda_2$ or $\lambda_2 > \lambda_1$ holds. Since the operator $\mathbf{M}$ is quasimonotonic, $\mathbf{M}\lambda_1 \neq \mathbf{M}\lambda_2$, which contradicts the condition of the theorem. The theorem is proved.

Below we will use either Theorem 1.1, or Lemma 1.2 and its variations.

1.2 INVERSE PROBLEMS FOR HYPERBOLIC EQUATIONS

First we will illustrate the application of Theorem 1.1 to the investigation of the uniqueness of the solution of the inverse problem for the linear hyperbolic equation, and then we proceed to more general cases.

Let $\mathbf{R}^3$ be an Euclidean space of (x,y), $x \in \mathbf{R}^2$, $-\infty < y < \infty$. Consider the problem: in the semispace $y \geq 0$ it is required to find a strictly positive function $\lambda(x,y) \subset C^1(\mathbf{R}^3)$, even in y, if

1. In the domain $x \in \mathbf{R}^2$, $y \in \mathbf{R}^1$, $0 \leq t \leq t_0(x,y)$, $t_0 > 0$ there exists a unique twice differentiable solution $u(x,y,t)$ of the Cauchy problem

$$\frac{\partial^2 u}{\partial t^2} = \Delta u + \lambda(x,y)u + f(x,y,t), \quad u|_{t=0} = \left.\frac{\partial u}{\partial t}\right|_{t=0} = 0,$$

   where Δ is the Laplace operator in (x,y).

2. In the domain $y = 0$, $0 \leq t \leq g_0(x)$ the function $\tau(x,t) = u|_{y=0}$, $0 \leq t \leq g_0$, $g_0(x) > 0$ is given. The inverse problem is reduced to the investigation of the operator equation $\mathbf{M}\lambda = \tau(x,t)$, $\lambda \in C^1(E)$, $E = \{(x,y), y > 0\}$. As covering $\{\omega\}$ of the semispace $y > 0$ we take as above the set of all hemiballs $\omega(\xi,p)$: $\omega(\xi,p) = \{(x,y) : |x-\xi|^2 + y^2 < \rho^2, \quad y > 0\}, \quad \xi \in \mathbf{R}^2, \quad \rho > 0$.

Let $\omega^-(\xi,p) = \{(x,y) : |x-\xi|^2 + y^2 < \rho^2, \quad y < 0\}$.

Theorem 1.2. *If a function $f(x,y,t)$ is continuous and $f > 0$, then operator $\mathbf{M}$ is quasimonotonic with respect to covering $\{\omega\}$ of domain $y > 0$.*

PROOF. Let $\lambda_1(x,y) > \lambda_2(x,y) > 0$, $(x,y) \in \omega_0 = \omega(\xi_0, p_0)$. Denote by $u_1(x,y,t)$, $u_2(x,y,t)$ the solutions of the Cauchy problem corresponding to the functions $\lambda_1(x,y)$ and $\lambda_2(x,y)$. Let $\tilde{\lambda} = \lambda_1 - \lambda_2$, $\tilde{u} = u_1 - u_2$. In accordance with the Kirchhoff formula we have

$$\tilde{u}(x,y,t) = \frac{1}{4\pi} \int\limits_{r \leq t} \frac{\lambda_1(q)\tilde{u}(q,t-r)}{r} dq + \frac{1}{4\pi} \int\limits_{r \leq t} \frac{\tilde{\lambda}(q)u_2(q,t-r)}{r} dq, \tag{1.4}$$

where

$$r^2 = \sum_{i=1}^{2} (x_i - q_i)^2 + (y - q_3)^2, \quad dq = dq_1 dq_2 dq_3.$$

Note that, if

$$u(x,y,t) = F(x,y,t) + \frac{1}{4\pi} \int\limits_{r \leq t} \frac{b(q)u(q,t-r)}{r} dq,$$

and $F|_{t>0} > 0$, $b > 0$, then (in any case for sufficiently small $t > 0$) the inequality $u(x,y,t) > 0$ holds. This follows, for example, from the proof of the existence of the solution $u(x,y,t)$ by the method of successive approximations.

By the condition of the problem and prepositions we have $f(x,y,t) > 0$, $\lambda_1 - \lambda_2 > 0$, $(x,y) \in \omega_0$, $\lambda_i > 0$. If $\lambda_i(x,y) = \lambda_i(x,-y)$ then $\lambda_1 - \lambda_2 > 0$, $(x,y) \in \omega_0^-$. Let t be such that the points q and (x,y) in (1.4) belong to $\omega^- \cup \omega_0$. From this using the note above we conclude that by the strict positivity of the second term in the right part of (1.4) for sufficiently small $t > 0$ we have $\tilde{u}(x,y,t) > 0$, $(x,y) \in \overline{\omega_0 \cup \omega^-}$. If $y = 0$ in the last inequality, we obtain $\tilde{u}(x,0,t) > 0$. This means operator $\mathbf{M}$ is quasimonotonic. The theorem is proved.

Let, as above, $\mathbf{R}^3$ be a three-dimensional Euclidean space for (x,y), $x \in \mathbf{R}^2$, $-\infty < y < \infty$ and $\Delta = \dfrac{\partial^2}{\partial y^2} + \sum\limits_{i=1}^{2} \dfrac{\partial^2}{\partial x_i^2}$ is the Laplace operator.

Denote by $\{a\}$ a set of infinite differentiable functions $a(x,y)$ to be even and quasianalytical in y. Let $\lambda(x,y)$ and $\mu(x,y)$ be some functions belonging to set $\{a\}$, and $u(x,y,t)$ a solution of the Cauchy problem, $\lambda > 0$, $\mu > 0$,

$$\begin{gathered} \frac{1}{\lambda(x,y)} \cdot \frac{\partial^2 u}{\partial t^2} = \Delta u + F(x,y,t,\mu(x,y),u) + f(x,y,t), \\ u\big|_{t=0} = \left.\frac{\partial u}{\partial t}\right|_{t=0} = 0, \end{gathered} \tag{1.5}$$

where F and f are fixed infinite differentiable functions. It is well known that a solution $u(x,y,t)$ of such a problem for the more general case exists and is unique in some domain of (x,y,t).

Here is considered the inverse problem: it is required to find one of the functions either $\lambda(x,y)$ or $\mu(x,y)$, if in the domain $|x| < r$, $0 \le t < g_0$ the function $\varphi(x,y) = u\big|_{y=0}$ is known, where $u(x,y,t)$ is the solution of the problem (1.5). Here the inverse problem is also reduced to the investigation of the equations $\mathbf{M}\lambda = \varphi(x,y)$, $\mathbf{M}\mu = \varphi(x,y)$.

Theorem 1.3. *Let function $F(x,y,t,\mu,u)$ increase strictly monotonic in variables μ, u in the domain $\mu \ge 0$, $u \ge 0$ and $F(x,y,t,0,0) = 0$, and the function $f(x,y,t) > 0$. Then the inverse problem has no more than one solution in the class $\{a\}$.*

PROOF. For the proof of the theorem it suffices to verify the operators $\mathbf{M}\lambda$ and $\mathbf{M}\mu$ being quasimonotonic. At first we note, that for every $\lambda(x,y) \in \{a\}$ and $\mu(x,y) \in \{a\}$ solution $u(x,y,t)$ of problem (1.5) for sufficiently small t must satisfy the inequality $u(x,y,t) > 0$, $t > 0$.

Really, in accordance to Sobolev's formula (see (Sobolev, 1963)) we have

$$u = \frac{1}{4\pi} \int\limits_{r \le t} [f]\sigma dv + \frac{1}{4\pi} \int\limits_{r \le t} [F]\sigma dv + \frac{1}{4\pi} \int\limits_{r \le t} [u]\Delta\sigma dv.$$

(Here we use the meanings of (Sobolev, 1963))
Since σ and $\Delta\sigma$ have an order of increasing no more than $1/\tau$ and $\sigma > 0$, it is clear by the formula, that the sign of u is defined by the first term of the right part of this formula. In particular, if $f(x,y,t) > 0$, then $u(x,y,t) > 0$.

Denote as ω^+ and ω^- open hemiballs of radius $\rho > 0$ centered at point $\xi \in \mathbf{R}^2$, $|\xi| < r$,

$$\omega^+(\xi,p) = \{(x,y), y > 0 : |x-\xi|^2 + y^2 < \rho^2\},$$

$$\omega^-(\xi,p) = \{(x,y), y < 0 : |x-\xi|^2 + y^2 < \rho^2\}.$$

Let $\mu_1(x,y)$ and $\mu_2(x,y)$ be two solutions of the problem ($\lambda(x,y)$ is fixed) and let $\mu_1(x,y) > \mu_2(x,y)$, $(x,y) \in \omega = \omega^+ \cup \omega^-$. Denote by $u_1(x,y,t)$, $u_2(x,y,t)$ solutions of the Cauchy problem

$$\frac{1}{\lambda} \cdot \frac{\partial^2 u_i}{\partial t^2} = \Delta u_i + F(x,y,t,\mu_i,u_i) + f, \quad i = 1,2,$$

$$u_i\big|_{t=0} = \frac{\partial u_i}{\partial t}\bigg|_{t=0} = 0,$$

corresponding to the functions $\mu_1(x,y)$ and $\mu_2(x,y)$. Let $w = u_1 - u_2$. Then we have

$$\frac{1}{\lambda} \cdot \frac{\partial^2 w}{\partial t^2} = \Delta w + F(x,y,t,\mu_1,w+u_2) - F(x,y,t,\mu_2,u_2),$$

$$w\big|_{t=0} = \frac{\partial w}{\partial t}\bigg|_{t=0} = 0.$$

Since $\mu_1(x,y) > \mu_2(x,y)$, $(x,y) \in \omega$ and as above $u_2(x,y,t) > 0$, by F being monotonic the inequality $F(x,y,t,\mu_1,u_2) > F(x,y,t,\mu_2,u_2)$ holds. So by using, for example, Sobolev's formula, we obtain that $(x,y) \in \bar{\omega}$, $w(x,y,t) > 0$ at least for small t. Setting $y = 0$ in the last inequality we have $\varphi_1(x,t) - \varphi_2(x,t) = w(x,0,t) > 0$, this is in conflict with the problem condition.

Let now $\mu(x,y)$ be fixed and let $\lambda_1(x,y)$, $\lambda_2(x,y)$ be two different solutions of the inverse problem. As above, let $w = u_1 - u_2$ and $\lambda_1(x,y) > \lambda_2(x,y)$, $(x,y) \in \omega$. Now we have

$$\frac{1}{\lambda_1} \cdot \frac{\partial^2 w}{\partial t^2} = \Delta w + F(x,y,t,\mu_1,w+u_2) - F(x,y,t,\mu_2,u_2) + \left(\frac{1}{\lambda_2} - \frac{1}{\lambda_1}\right)\frac{\partial^2 u_2}{\partial t^2},$$

$$w\big|_{t=0} = \frac{\partial w}{\partial t}\bigg|_{t=0} = 0.$$

By the condition of the theorem $f(x,y,t) > 0$, as is $\dfrac{\partial^2 u_2}{\partial t^2}$. Therefore, by inequality $\lambda_1(x,y) > \lambda_2(x,y)$, $(x,y) \in \omega$ we have

$$\left(\frac{1}{\lambda_2} - \frac{1}{\lambda_1}\right)\frac{\partial^2 u_2}{\partial t^2} > 0, \quad (x,y) \in \omega,$$

which as above is in conflict with $w(x,0,t) > 0$. By Theorem 1.1 and the obtained results the statement of the theorem follows. For reduction to other problems see (Lavrent'ev, 1974).

1.3 MULTIDIMENSIONAL INVERSE KINEMATIC PROBLEM OF SEISMICS

Let $\mathbf{R}^{n+1}$ be the $(n+1)-$ dimensional Euclidean space for (x,y), $x \in \mathbf{R}^n$, $-\infty < y < \infty$, $\mathbf{n}$ — radius vector of the sphere $\Omega = \{x, |x| = 1\}$.

Consider the problem: in semispace $y \geq 0$ it is required to find an infinite differentiable function $\lambda(x,y) \in \{\lambda\}$, $0 < \lambda < \mathbf{M} < \infty$ such that:

1. The shortest line $\gamma(a,b,\mathbf{n}) \subset \bar{E}$ exists of metrics $\mathrm{d}s^2 = \lambda^2(x,y)(\mathrm{d}x^2 + \mathrm{d}y^2)$ joining points $a\mathbf{n}$, $b\mathbf{n}$, $-\infty < a < \infty$, $-\infty < b < \infty$, $\mathbf{n} \in \Omega$, and

2. In the domain $|a| < \infty$, $|b| < \infty$, $\mathbf{n} \in \Omega$ the function

$$\tau(a,b,\mathbf{n}) = \int\limits_{\gamma(a,b,\mathbf{n})} \lambda(x,y)\sqrt{\mathrm{d}x^2 + \mathrm{d}y^2} \equiv M\lambda$$

is given.

The problem of determining the function $\lambda(x,y)$ by $\tau(a,b,\mathbf{n})$ is called the inverse kinematic problem of seismics, where $\lambda(x,y)$ is interpretated as the inverse of the velocity $\lambda = 1/v(x,y)$ and $\tau(a,b,\mathbf{n})$ is the time of moving of the perturbation by $\gamma(a,b,\mathbf{n})$.

Let

$$\lambda_0 = \lambda(x,0) = \lim_{\substack{a\to b=t \\ t\mathbf{n}=x}} \frac{\tau(a,b,\mathbf{n})}{|b-a|}$$

In $\mathbf{R}^n$ we consider metric $\mathrm{d}s_0^2 = \lambda_0^2(x)\mathrm{d}x^2 = \lambda_0^2(x)\,|\mathrm{d}x|^2$. Denote by $\gamma^0(a,b,\mathbf{n})$ the shortest line of metric $\mathrm{d}s_0$ joining points $a\mathbf{n}$, $b\mathbf{n}$. Denote by

$$\tau^0(a,b,\mathbf{n}) = \int\limits_{\gamma^0(a,b,\mathbf{n})} \lambda_0\,|\mathrm{d}x|\,.$$

Theorem 1.4. *If $\tau(a,b,\mathbf{n}) \neq \tau^0(a,b,\mathbf{n})$ for every a, b and $\mathbf{n}$, then the inverse kinematic problem of seismic has no more than one quasianalitical solution in y: $\lambda(x,y) \in \{\lambda\}$.*

PROOF. Show that under condition of the theorem the operator

$$\mathbf{M}\lambda = \int\limits_{\gamma(a,b,\mathbf{n})} \lambda(x,y)\sqrt{\mathrm{d}x^2 + \mathrm{d}y^2}, \quad \lambda(x,y) \in \{\lambda\}_E$$

is quasimonotonic with respect to the covering $\{\omega\}$ of domain $E = \{(x,y) : y > 0\}$. In the same way as Theorem 1.1 the statement will be proved. Let $\omega_0 = \omega(\xi_0,t)$ be a hemiball of radius $t > 0$ centered at $\xi_0 \in \mathbf{R}^n$ belonging to the semispace, and let $\lambda_1(x,y) > \lambda_2(x,y)$, $(x,y) \in \omega_0$. Denote by γ_i the shortest line of metric $\mathrm{d}s_i^2 = \lambda_i^2(\mathrm{d}x^2 + \mathrm{d}y^2)$, such that $\gamma_i \subset \omega_0$.

By the condition of the theorem the shortest line γ_i cannot belong to the hyperplane $y = 0$ totally, in contrast

$$\tau_i = \int\limits_{\gamma_i} \lambda_i(x,y)\sqrt{\mathrm{d}x^2 + \mathrm{d}y^2} = \int\limits_{\gamma_i^0} \lambda_i(x,0)\,|\mathrm{d}x| = \tau_i^0.$$

Using the inequality $\lambda_1(x,y) > \lambda_2(x,y)$, $(x,y) \in \omega_0$ and the last remark, we have

$$\tau_1 = \int\limits_{\gamma_1} \lambda_1(x,y)\sqrt{\mathrm{d}x^2 + \mathrm{d}y^2} > \int\limits_{\gamma_1} \lambda_2(x,y)\sqrt{\mathrm{d}x^2 + \mathrm{d}y^2} \geq \int\limits_{\gamma_2} \lambda_2(x,y)\sqrt{\mathrm{d}x^2 + \mathrm{d}y^2} = \tau_2,$$

which means operator $\mathbf{M}$ is quasimonotonic. The theorem is proved.

More genéral results are included in the papers by (Anikonov, 1971; 1975).

REMARK. In the plane case, instead of the condition of Theorem 1.4 it may be taken as $\tau(a,b) < \tau(a,b_1) + \tau(b_1,b)$, $a < b_1 < b$. By the fulfilment of this inequality every interior point on the shortest line $\gamma(a,b)$ of metrics $\mathrm{d}s^2 = \lambda^2(x,y)(\mathrm{d}x^2 + \mathrm{d}y^2)$ cannot belong to the straight $y = 0$.

For $n \geq 2$ variables a, b can be changed in the domain $|a| < a_0$, $|b| < b_0$, where $a_0 > 0$, $b_0 > 0$ are constant.

1.4 ON THE UNIQUENESS OF THE SOLUTION OF THE FREDHOLM AND VOLTERRA FIRST KIND INTEGRAL EQUATIONS

As we noted above, inverse problems for differential equations reduce to the first kind operator equations, and often to the first kind integral equations. Let us consider multidimensional integral equations and prove the uniqueness theorems.

Let $\mathbf{R}^{n+1}$ be the Euclidean space for (x, y), $x \in \mathbf{R}^n$, $-\infty < y < \infty$, $n \geq 0$, let V be a compact metric set of v. Denote by $\omega(\xi, t)$, $\xi \in \mathbf{R}^n$, $t > 0$, $(n+1)-$ dimensional open set with properties:

1. $\omega(\xi, t)$ is contained in the semispace $y > 0$ and unicity $\{\omega\}$ is the covering of this space.

2. $\omega(\xi, t)$ depends continuously of ξ and t, and $\lim\limits_{t\to 0} \omega(\xi, t) = (\xi, 0)$.

An example of the set $\omega(\xi, t)$ may be a hemiball of radius $t > 0$ centered at point $\xi \in \mathbf{R}^n$

$$\omega(\xi, t) = \left\{(x, y): \quad |x - \xi|^2 + y^2 < t^2, \quad \text{and } y > 0\right\}.$$

Let $M(x, y, \xi, t, v)$ be a real continuous function on $\mathbf{R}^{n+1} \times \mathbf{R}^{n+1} \times V$ possessing properties

1. $M(x, y, \xi, t, v) > 0$, $(x, y) \in \omega(\xi, t)$,

2. $M(x, y, \xi, t, v) \equiv 0$, $(x, y) \in \mathbf{R}^{n+1} \backslash \bar{\omega}(\xi, t)$,

where $\bar{\omega}(\xi, t)$ is closure of $\omega(\xi, t)$.

Let us consider this equation with respect to the function $\lambda(x, y) \in \{\lambda\}$:

$$\mathbf{M}\lambda \equiv \inf_v \int\limits_{\omega(\xi,t)} M(x, y, \xi, t, v)\lambda(x, y)\mathrm{d}x_1 \ldots \mathrm{d}x_n \mathrm{d}y = w(\xi, t). \tag{1.6}$$

Note, that if function M does not depend on v, then the considered equation is the multidimensional analog of Volterra equation. In particular, if $n = 0$ then $\mathbf{M}\lambda = \int\limits_0^t M(y, t)\lambda(y)\mathrm{d}y$. Equation $\varphi(t) = \int\limits_0^t M(y, t)\lambda(y)\mathrm{d}y$ is the first kind Volterra equation.

Theorem 1.5. *If* $\mathbf{M}\lambda_1 = \mathbf{M}\lambda_2$, $\lambda_i \in \{\lambda\}$, $i = 1, 2$, *then* $\lambda_1(x, y) = \lambda_2(x, y)$, $x \in \mathbf{R}^n$, $0 \leq y < \infty$.

PROOF. It is sufficiently to state operator $\mathbf{M}$ being quasimonotonic, defined by (1.6), $\lambda \in C$.

As covering $\{\omega\}$ of the semispace $y > 0$ we understand sets $\omega(\xi, t)$, satisfy conditions 1 and 2.

Let $\lambda_1(x, y) > \lambda_2(x, y)$, $(x, y) \in \omega_0 = \omega(\xi_0, t_0)$, $\lambda_i \in \{\lambda\}$, $i = 1, 2$. Since the kernel $M(x, y, \xi, t, v)$ is continuous and has properties 1 and 2, functions

$$w_i(\xi, t, v) = \int\limits_{\omega(\xi,t)} M(x, y, \xi, t, v)\lambda_i(x, y)\mathrm{d}x\mathrm{d}y, \quad i = 1, 2,$$

are continuous in the domain $\xi \in \mathbf{R}^n$, $t > 0$, $v \in V$. Thus as V is a compact metric space there exist elements $v_i \in V$ so that

$$w_i(\xi, t) = \inf_v w_i(\xi, t, v) = w_i(\xi, t, v_i), \quad i = 1, 2.$$

By the above we have

$$\begin{aligned} w_1(\xi_0, t_0) &= \int_{\omega_0} M(x, y, \xi_0, t_0, v_1)\lambda_1(x, y)dxdy > \int_{\omega_0} M(x, y, \xi_0, t_0, v_1)\lambda_2(x, y)dxdy \\ &\geq \int_{\omega_0} M(x, y, \xi_0, t_0, v_2)\lambda_2(x, y)dxdy = w_2(\xi_0, t_0) \end{aligned}$$

which means operator $\mathbf{M}$ is quasimonotonic. The theorem is proved.

Suppose now that sets $\omega(\xi, t)$ have additional properties:

1. For every $h \geq 0$ set $\omega(\xi, t, h) = \omega(\xi, t) \cap \{y > h, \quad x \in \mathbf{R}^n\}$, $t > h$, is an $(n+1)$-dimensional open set,

2. $\omega(\xi, t, h)$ depends continuously of ξ, t, h and $\lim_{t \to h} \omega(\xi, t, h) = (\xi, h)$.

Define a class $\{\lambda\}_\infty$ of $\lambda(x, y)$, given by the definition $\lambda(x, y) \in \{\lambda\}_\infty$, there exists some sequence of numbers h_i, $0 = h_0 \leq h_1 \leq \ldots \leq h_m \ldots$, and some sequence of functions $\lambda_i(x, y)$, $h_i \leq y \leq h_{i+1}$, $x \in \mathbf{R}^n$, analytical in y, which are infinite differentiable with respect to (x, y) such that $\lambda(x, y) = \lambda_i(x, y)$, $x \in \mathbf{R}^n$, $h_i \leq y \leq h_{i+1}$. In other words, class $\{\lambda\}_\infty$ consists of functions that are piecewise analytical in y and infinite differentiable with respect to (x, y).

Theorem 1.6. *If* $\mathbf{M}\lambda_1 = \mathbf{M}\lambda_2$, $\lambda_i \in \{\lambda\}_\infty$, $i = 1, 2$, *then* $\lambda_1(x, y) = \lambda_2(x, y)$, $x \in \mathbf{R}^n$, $y \geq 0$.

PROOF. Let $\lambda_1(x, y)$ and $\lambda_2(x, y)$ $\lambda_i \in \{\lambda\}_\infty$ be two solutions of equation (1.6), such as $\mathbf{M}\lambda_1 = \mathbf{M}\lambda_2$. By continuity of the kernel $M(x, y, \xi, t, v)$ and conditions 1 and 2, functions

$$w_i(\xi, t, v) = \int_{\omega(\xi, t)} M(x, y, \xi, t, v)\lambda_i(x, y)dxdy$$

are continuous.

Therefore v is a compact metric space, from this in particular it follows the existence of v_i so that

$$w_i(\xi, t) = w_i(\xi, t, v_i) = \inf_v w_i(\xi, t, v).$$

Suppose that $\lambda_1(x, y) \neq \lambda_2(x, y)$. From this and by the definition of class $\{\lambda\}_\infty$ it follows the existence of the interval $[a, b]$, $0 \leq a < b < \infty$ so that the following conditions are true:

1. $\lambda_1(x, y) = \lambda_2(x, y)$, $x \in \mathbf{R}^n$, $y < a$,

2. $\lambda_i(x, y)$ are analytical in y when $a \leq y < b$, and $\lambda_1(x, y) \neq \lambda_2(x, y)$, $x \in \mathbf{R}^n$, $a \leq y < b$.

Let $\omega^1(\xi, t)$ be the intersection of $\omega(\xi, t)$ with the semispace $y > 0$. By Lemma 1.2 and proposition $\lambda_1(x, y) \neq \lambda_2(x, y)$ there exists a set $\omega_0^1 = \omega^1(\xi_0, t_0) = \omega(\xi_0, t_0) \cap \{y > a\}$ so that

1. Set ω_0^1 belongs to the strip $a < y < b$, $x \in \mathbf{R}^n$,

2. $|\lambda_1(x,y) - \lambda_2(x,y)| > 0, \quad (x,y) \in \omega_0^1$.

Let $\lambda_1 > \lambda_2$, $(x,y) \in \omega_0^1$. Taken this into consideration and by $\lambda_1 = \lambda_2$, $x \in \mathbf{R}^n$, $y < a$ we have.

$$w_1(\xi_0,t_0) = \int\limits_{\omega(\xi_0,t_0)} M(x,y,\xi_0,t,v_1)\lambda_1 dxdy > \int\limits_{\omega(\xi_0,t_0)} M(x,y,\xi_0,t,v_1)\lambda_2 dxdy \geq w_2(\xi_0,t_0).$$

Thus the existence of the point is shown (ξ_0,t_0) with $w_1(\xi_0,t_0) > w_2(\xi_0,t_0)$. The last inequality is in conflict with the condition of the theorem. The theorem is proved.

Example of nonuniqueness of a solution of the Volterra integral equation with a positive kernel.

Let $\lambda(t) \not\equiv 0$, $t \geq 0$, $\lambda(0) = 0$ be an infinite differentiable function with zero in every interval $(0,\alpha)$, $\alpha > 0$, as for example $\lambda(t) = \exp(-1/t^2)\sin(1/t)$, $\lambda(0) = 0$. Show that there exists a kernel $K(x,t) > 0$, $x \neq 0$, $t \neq 0$ such that for every $x > 0$ the equality

$$\int\limits_0^x K(x,t)\lambda(t)dt = 0$$

holds.

Let $E^+(x)$, $E^-(x)$ be sets such that

$$E^+(x) = \{t : \lambda(t) > 0\} \cap (0,x),$$

$$E^-(x) = \{t : \lambda(t) \leq 0\} \cap (0,x)$$

Set

$$\mu^+(x,t) = \begin{cases} 1, & \text{if } t \in E^+(x), \\ 0, & \text{if } t \in E^-(x), \end{cases} \qquad 0 \leq t \leq x$$

$$\mu^-(x,t) = \begin{cases} -1, & \text{if } t \in E^-(x), \\ 0, & \text{if } t \notin E^-(x), \end{cases}$$

Introduce functions

$$g^+(x) = \int\limits_{E^+(x)} \lambda(t)dt, \quad g^-(x) = \int\limits_{E^-(x)} \lambda(t)dt.$$

Define kernel $K(x,t)$ by $K(x,t) = -\mu^+(x,t)g^-(x) - \mu^-(x,t)g^+(x)$. It is obvious, that $K(x,t) > 0$, $x \neq 0$. We have

$$\int\limits_0^x K(x,t)\lambda(t)dt = \int\limits_{E^+(x)} K(x,t)\lambda(t)dt + \int\limits_{E^-(x)} K(x,t)\lambda(t)dt = g^+(x)g^-(x) - g^+(x)g^-(x) = 0.$$

Thus the kernel $K(x,t)$ is constructed such as for every x, $x \geq 0$

$$\int\limits_0^x K(x,t)\lambda(t)dt = 0.$$

Note that if $\tilde{\lambda}(t)$ is a quasianalytical function in the domain $t \geq 0$ then from the equality

$$\int_0^x K(x,t)\tilde{\lambda}(t)dt = 0$$

by Theorem 1.1 it follows that $\tilde{\lambda}(t) = 0$.

On the uniqueness of solution of the Fredholm first kind integral equation

Here we consider the multidimensional first kind integral equation with singularity on the diagonal or with this properties after repeated differentiation. Let $\mathbf{R}^n$be the $n-$ dimensional Euclidean space for $x = (x_1, x_2, \ldots, x_n)$, and let Ω, ω be balls in $\mathbf{R}^n$ with:

$$\Omega = \{x : |x| < R\}, \quad \omega = \{x : |x| < r\}, \quad R > r > 0.$$

With ξ we will denote the points of sphere annulus $C = \Omega \backslash \bar{\omega}$, here the overbar, as usual, defines the closure of the set. Consider first the integral equation

$$\int_\omega M(x,\xi)\lambda(x)dx = u(\xi), \quad dx = dx_1 dx_2 \ldots dx_n, \tag{1.7}$$

where kernel $M(x,\xi)$ has properties:

1. $M(x,\xi)$ is given for all $x \in \mathbf{R}^n$, $\xi \in \mathbf{R}^n$, continuous for $\xi \neq x$.

2. For every $q \geq 0$ equality

$$\lim_{|x-\xi|\to 0} M(x,\xi)(x-\xi)^q = \infty$$

holds.

As example of such kernels we can take $M(x,\xi) = \exp(\varepsilon/|x-\xi|)\,|x-\xi|^\delta$, where $\varepsilon > 0$, δ can be any number. For $\varepsilon = 0$ such kernels take place in inverse problems of potential.

Let $\{\lambda\}_a$ be a set of functions $\lambda(x)$, $x \in \bar{\omega}$, presented by $\lambda(x) = a(x)g(x) + b(x)$ where $a(x)$ is some analytical function in closed ball $\bar{\omega}$, and $g(x)$, $b(x)$ are fixed measurable functions, so that function $g(x)$ satisfies the inequalities $0 < M_0 \leq g(x) \leq M_1$.

Note that if $g(x) = 1$, $b(x) = 0$ then set of function $\{\lambda\}_a$ includes the set of all analytical functions in $\bar{\omega}$.

Theorem 1.7. *Equation (1.7) has no more than one solution $\lambda(x) \in \{\lambda\}_a$.*

PROOF. Suppose, that equation (1.7) has two solutions $\lambda_1(x)$ and $\lambda_2(x)$ $\lambda_1(x) \neq \lambda_2(x)$, $\lambda(x) \in \{\lambda\}_a$, $i = 1,2$. Denote as $a(x) = a_1(x) - a_2(x)$. We have $\int_\omega M(x,\xi)g(x)a(x)dx = 0$. Show, that if $a(x) \neq 0$ then there exists at least one point $\xi_0 \in C$ so that $\int_\omega M(x,\xi_0)g(x)a(x)dx \neq 0$. By this the statement of the theorem is be established. Let s be a point on the sphere $S = \{x : |x| = r\}$, $B(s,t)$ is an $n-$ dimensional open ball of radius $t > 0$ centered at point $s \in S$. Denote by $\omega(s,t)$ and $C(s,t)$ the intersection $B(s,t) \cap \omega$, $B(s,t) \cap C$.

Show at first that there exists point $s_0 \in S$ and numbers t_0 and t_1, $t_0 > t_1$, so that the inequalities hold

1. $|a(x)| \geq A(r-|x|)^{\alpha}$, $x \in \bar{\omega}(s_0, t_0) = \bar{\omega}_0$, where $A > 0$, $\alpha > 0$ are some constants,

2. $M(x,\xi)g(x)a(x) \geq k_0$, $x \in \bar{\omega}\backslash\bar{\omega}_0$, $\xi \in C(s_0, t_0)$, k_0 is some constant.

Inequality 1 follows from Lemma 1.2, and estimate 2 follows from the continuity of functions $M(x,\xi)$, and $a(x)$, and $g(x)$.

Note that as $a(x) = a_1(x) - a_2(x)$ the inequality $a(x) \geq A(r-|x|)^{\alpha}$, $x \in \omega_0$ is true, in another case functions a_1, a_2 may be changed. Then by the condition the kernel $M(x,\xi)$ is continuous in domain $\xi \neq x$ and when $|x-\xi| \to 0$, $M(x,\xi) \to \infty$. Therefore for every $s \in S$ it exists intersection $\omega(s,t)$ and $C(s,t)$ so that $M(x,\xi) > 0$ when $x \in \omega(s,t)$, $\xi \in C(s,t)$. By the above we include there exist intersections $\omega_0 = \omega(s_0, t_0)$, $C_0 = C(s_0, t_0)$ with

1. $M(x,\xi) > 0$, $x \in \omega_0$, $\xi \in C_0$,

2. $a(x) \geq A(r-|x|)^{a}$, $x \in \omega_0$,

3. 3. $M(x,\xi)g(x)a(x) \geq k_0$, $\xi \in C_0$, $x \in \bar{\omega}\backslash\bar{\omega}_0$.

Using estimates we have

$$\begin{aligned}\int_{\omega} M(x,\xi)g(x)a(x)\mathrm{d}x &= \int_{\omega\backslash\omega_0} M(x,\xi)g(x)a(x)\mathrm{d}x + \int_{\omega_0} M(x,\xi)g(x)a(x)\mathrm{d}x \\ &\geq k_0 v_0 + M_0 A \int_{\omega_0} M(x,\xi)(r-|x|)^{\alpha}\mathrm{d}x,\end{aligned}$$

where v_0 is volume of domain $\omega\backslash\omega_0$. Since $\lim\limits_{x\to\xi} M(x,\xi)\,|x-\xi|^{q} = \infty$, $\int\limits_{\omega_0} M(x,\xi)(r-|x|)^{\alpha}\mathrm{d}x \to \infty$, when $\xi \to s_0$. Thus there exists a point ξ_0 with $\int\limits_{\omega} M(x,\xi)g(x)a(x)\mathrm{d}x > 0$. The theorem is proved.

Suppose now that kernel $M(x,\xi)$ is indefinite differentiable in the domain $\xi \neq x$. Denote as D^{β} a differential operator with respect to $\xi = (\xi_1, \xi_2, \dots, \xi_n)$:

$$D^{\beta} = \frac{\partial^{\beta_1+\beta_2+\ldots+\beta_n}}{\partial\xi_1^{\beta_1}\partial\xi_2^{\beta_2}\ldots\partial\xi_n^{\beta_n}}.$$

Let $M_{\beta}(x,\xi) = D^{\beta}M(x,\xi)$. Let kernels $M_{\beta}(x,\xi)$ possess properties:

1. $M_{\beta}(x,\xi)$ satisfy condition 1, above,

2. There exists a sequence of numbers $\{q_m\}$, $q_m \to \infty$ so that

$$\lim_{x\to\xi} M_{\beta}(x,\xi)\,|x-\xi|^{q_m} = \infty, \quad |\beta| = m.$$

Consider the following equation with such kernel

$$\int_{\omega} M(x,\xi)\lambda(x)\mathrm{d}x = u(\xi). \tag{1.8}$$

Theorem 1.8. *Equation (1.8) has no more than one solution $\lambda(x) \in \{\lambda\}_a$.*

Proof. If $\lambda_i = a_i(x)g(x) + b(x)$, $i = 1,2$ are two nonequal solutions of (1.8), then the equalities

$$\int_\omega M_\beta(x,\xi)a(x)g(x)dx = 0, \quad m = 0,1,2,\dots \tag{1.9}$$

hold, where $a(x) = a_1(x) - a_2(x)$, $M_\beta(x,\xi) = D^\beta M(x,\xi)$, $|\beta| = m$. As in the proof of Theorem 1.1 estimates state

1. $M_\beta(x,\xi) > 0$, $x \in \omega_0^m$, $\xi \in C_0^m$,

2. $a(x) \geq A(r - |x|)^\alpha$, $x \in \omega_0^m$,

3. $M_\beta(x,\xi)g(x)a(x) \geq k_m$, $\xi \in C_0^m$, $x \in \bar{\omega}\backslash\bar{\omega}_0^m$.

From this and the inequality above follows

$$\int_\omega M_\beta(x,\xi)agdx \geq k_m v_0^m + AM_0 \int_{\omega_0^m} M_\beta(x,\xi)(r - |x|)^\alpha.$$

If the number m is sufficiently large, then $\int_{\omega_0} M_\beta(x,\xi)(r - |x|)^\alpha dx$, where $\xi \to s_0$ tends to infinity. Thus there exists a point ξ_0 such that $\int_\omega M_\beta(x,\xi_0)a(x)g(x)dx = 0$, $|\beta| = m$, which is in conflict with (1.9). The theorem is proved.

The example of nonuniqueness of the solution.

Let the kernel $M(x,\xi) = 1$, where $|\xi| > r$, $|x| < r/2$ and let $\lambda(x) \not\equiv 0$ be an infinite differentiable function in circle $\omega = \{x : |x| \leq r\}$ so that

1. $\lambda(x) \equiv 0$, $x \in \{x : |x| > r/2\}$,

2. $\int_{|x|<r/2} \lambda(x)dx = 0$.

It is obvious that $\int_\omega M(x,\xi)\lambda(x)dx = 0$.

1.5 ON THE UNIQUENESS OF THE SOLUTION OF INTEGRAL EQUATIONS OF THE FIRST KIND WITH ENTIRE KERNEL

In this section we consider the multidimensional integral equation of the first kind

$$w(x) = \int_{\mathbf{R}^n} k(x,y)\lambda(y)dy, \quad x \in D, \tag{1.10}$$

where D is a domain in the real Euclidean space $\mathbf{R}^n$, $n \geq 1$, $k(x,y)$ is a complex-valued kernel in $\mathbf{R}^n \times \mathbf{R}^n$, represented in the form

$$k(x,y) = \int_B \varphi(p,q)\, e^{i(px+qy)}dpdq. \tag{1.11}$$

Here an integrand function $\varphi(p,q)$ is complex-valued, continuous in $\mathbf{R}^n \times \mathbf{R}^n$, finite with compact support $\overline{B}$.

It is known (Ronkin, 1974), that the kernel $k(x,y)$ is continue on an entire analytic function of exponential type. In the present section we give conditions to domain B and function $\varphi(x,y)$, guaranteeing the uniqueness of solution $\lambda(y)$ of equation (1.10) in the class of continuous complex-valued finite functions.

Let $\alpha(t) > 0$, $\alpha(0) = 0$ be a continuous function in $\mathbf{R}^1$, $\varepsilon > 0$, $\delta > 0$ be fixed numbers and let space $R^n(p)$ be defined in the following way

$$R^n(p) = \{q : (p,q), \quad q \in \mathbf{R}^n\}.$$

Consider set

$$\omega(p) = \left\{q : \sum_{i=1}^{n-1} (p_i - q_i)^2 + q_n^2 < \alpha(p_n), \quad q_n > 0\right\}.$$

The set $\omega(p)$ is an open hemiball in $R^n(p)$ of radius $\alpha(p_n)$ centered at $p_0 = (p_1, p_2, \dots, p_{n-1}, 0)$. Further we propose that domain B and function $\varphi(p,q) = \varphi_1(p,q) + \mathrm{i}\varphi_2(p,q)$ have properties:

1. For any p, $0 < p_i < \varepsilon$, $i = 1, 2, \dots, n$, intersection $B(p) = B \cap R(p)$ is non-empty open set in $R^n(p)$ and $B(p) \subset \omega(p)$,

2. In a domain $\tilde{B} = B \cap \left\{(p,q) : \sum_{i=1}^{n-1} p_i^2 + q_i^2 < \delta\right\}$ the following inequalities hold:

$$\varphi_1(p,q) \geq 0, \quad \varphi_2(p,q) \geq 0, \quad \varphi_1(p,q) + \varphi_2(p,q) > 0.$$

In the case where $n = 1$ as example of a domain B satisfying condition 1 may be the domain with

$$B = \{(p_1, q_1) : 0 < q_1 < \alpha(p_1), \quad 0 < p_1 < 1\},$$

where $\alpha(t) > 0$, $\alpha(0) = 0$, $0 \leq t \leq 1$ is a continuous function. An example of domain B, not satisfying condition 1 is the square

$$B = \{(p_1, q_1) : 0 < p_1 < 1, \quad 0 < q_1 < 1\}.$$

Theorem 1.9. *Let conditions 1 and 2 hold. Then equation (1.10) has no more than one continuous finite solution $\lambda(y)$.*

PROOF. It is clear from (1.10) and (1.11) that for any continuous finite solution $\lambda(y)$ of equation (1.10), function $w(x)$ is an entire analytic function. Therefore, if $w(x) = 0$, $x \in D$, then $w(x) = 0$, $x \in \mathbf{R}^n$. Let $w(x) = 0$, $x \in \mathbf{R}^n$. Show, that $\lambda(y) = 0$, $y \in \mathbf{R}^n$.

We have

$$\int_{\mathbf{R}^n} k(x,y)\lambda(y)\mathrm{d}y = \int_{\mathbf{R}^n} \left[\int_B \varphi(p,q)\,\mathrm{e}^{\,\mathrm{i}(px+qy)}\mathrm{d}p\mathrm{d}q\right] \lambda(y)\mathrm{d}y = 0.$$

Changing the order of integration we obtain the equality

$$\int_B \left[\varphi(p,q)\,\mathrm{e}^{\,\mathrm{i}(px+qy)} \int_{\mathbf{R}^n} \mathrm{e}^{\,\mathrm{i}qy}\lambda(y)\mathrm{d}y\right] \mathrm{d}p\mathrm{d}q = 0. \tag{1.12}$$

Let

$$\tilde{\lambda}(q) = \int\limits_{\mathbf{R}^n} e^{iqy} \lambda(y) dy. \tag{1.13}$$

By such designation equality (1.12) acquires the form

$$\int\limits_{B} \varphi(p,q)\, e^{ipx} \tilde{\lambda}(q) dp dq = 0.$$

Passing on to an iterated integral we come to the relation

$$\int\limits_{\mathbf{R}^n} \left[\int\limits_{B(p)} \varphi(p,q) \tilde{\lambda}(q) dq \right] e^{ipx} dp = 0. \tag{1.14}$$

Here $B(p)$, as above, is the intersection of the domain B by the plane $R^n(p)$ of variables q of dimension n, passing through p and orthogonal to space $\mathbf{R}^n$ of variables p. Let

$$\tilde{\tilde{\lambda}}(p) = \int\limits_{B(p)} \varphi(p,q) \tilde{\lambda}(q) dq.$$

In this case we rewrite equality (1.14) as

$$\int\limits_{\mathbf{R}^n} \tilde{\tilde{\lambda}}(p)\, e^{ipx} dp = 0.$$

Thus the Fourier transform of continuous function $\tilde{\tilde{\lambda}}(p)$ is equal to zero. Therefore $\tilde{\tilde{\lambda}}(p) = 0$, $p \in \mathbf{R}^n$ and we have

$$\int\limits_{B(p)} \varphi(p,q) \tilde{\lambda}(q) dq = 0. \tag{1.15}$$

Because under the hypothesis of the theorem the sought solution $\lambda(y)$ is finite continuous function, then it follows from (1.13) and by the Paley-Wiener theorem (Ronkin, 1974) that $\tilde{\lambda}(q) = \tilde{\lambda}_1(q) + i\tilde{\lambda}_2(q)$ is an entire analytic function.

Therefore, real functions $\tilde{\lambda}_i(q)$, $i = 1, 2$, are entire analytic functions too.

Determining the imaginary and real parts in (1.15) we obtain equation system

$$\begin{aligned} &\int\limits_{B(p)} (\varphi_1 \tilde{\lambda}_1 - \varphi_2 \tilde{\lambda}_2) dq = 0, \\ &\int\limits_{B(p)} (\varphi_1 \tilde{\lambda}_2 + \varphi_2 \tilde{\lambda}_1) dq = 0, \end{aligned} \tag{1.16}$$

Show that $\tilde{\lambda}_1(q) = 0$, $\tilde{\lambda}_2(q) = 0$, $q \in \mathbf{R}^n$. From here by virtue of (1.13) equality $\lambda(y) = 0$, $y \in \mathbf{R}^n$ follows. We first prove the statement concerning the analytic functions.

Let $p_0 = (p_1^0, p_2^0, \ldots, p_{n-1}^0, 0)$ be a fixed point, $r > 0$ be a fixed number and ω be an open hemiball of radius $r > 0$ centered at p_0:

$$\omega_0 = \left\{ q : \sum_{i=1}^{n-1} (p_i - q_i^2) < r, \quad q_n > 0 \right\}.$$

Denote as $\{\omega\}$ the set of all open hemiballs centered at $(p_1, p_2, \ldots, p_{n-1}, 0)$ and such, that $\omega < \omega_0$.

If $f(q)$ is a real entire analytic function, then there exists an open hemiball $\tilde{\omega} \in \{\omega\}$ such, that for all $q \in \tilde{\omega}$ one of two inequalities holds: either $f(q) > 0$, or $f(q) < 0$.

Suppose, that this is not so. Then in every hemiball $\omega \in \{\omega\}$ there exists a point q^{ω} such, that $f(q^{\omega}) = 0$. In particular equality follows from this

$$f(q)\big|_{q_n=0} = 0, \quad |q| < r.$$

Because $f(q^{\omega}) = 0$ and $f(q_0^{\omega}) = 0$, where $q_0^{\omega} = (q_1^{\omega}, q_2^{\omega}, \ldots, q_{n-1}^{\omega}, 0)$ then in every hemiball there exists a point $\tilde{q}^{\omega}$ such, that

$$\left.\frac{\partial f}{\partial q_n}\right|_{q=\tilde{q}^{\omega}} = 0, \quad \tilde{q}^{\omega} \in \omega \in \{\omega\}.$$

This leads as above to equality $\left.\dfrac{\partial f}{\partial q_n}\right|_{q_n=0} = 0, \quad |q| < r.$

Hence, for any m, $m \geq 0$,

$$\left.\frac{\partial^m f}{\partial q_m^m}\right|_{q_n=0} = 0, \quad |q| < r,$$

which contradicts the relation $f(q) \not\equiv 0$. The statement is proved.

If we suppose now that, although one of entire functions $\tilde{\lambda}_1(q)$, $\tilde{\lambda}_2(q)$, satisfying equation system (1.16), is not identically equal to zero, then by virtue of the statement proved above, there exists $\tilde{\varepsilon} > 0$, $\tilde{\varepsilon} < \varepsilon$ and a point p such, that for any

$$q \in \omega(p) = \left\{ q : \sum_{i=1}^{n-1} (p_i - q_i)^2 + q_n^2 < \alpha(p_n), \quad q_n > 0 \right\}, \quad 0 < p_i < \tilde{\varepsilon}, \quad i = 1, 2, \ldots, n,$$

one of four relation holds

1. $\tilde{\lambda}_1(q) \geq 0$, $\tilde{\lambda}_2(q) \geq 0$, $\tilde{\lambda}_1(q) + \tilde{\lambda}_2(q) > 0$,

2. $\tilde{\lambda}_1(q) \leq 0$, $\tilde{\lambda}_2(q) \leq 0$, $\tilde{\lambda}_1(q) + \tilde{\lambda}_2(q) < 0$,

3. $\tilde{\lambda}_1(q) \leq 0$, $\tilde{\lambda}_2(q) \geq 0$, $\tilde{\lambda}_1(q) - \tilde{\lambda}_2(q) < 0$,

4. $\tilde{\lambda}_1(q) \geq 0$, $\tilde{\lambda}_2(q) \leq 0$, $\tilde{\lambda}_1(q) - \tilde{\lambda}_2(q) > 0$.

Under the hypotheses of the theorem at $\tilde{\varepsilon} < \delta$ for all (p, q)

$$(p, q) \in B \cap \left\{ (p, q) : \sum_{1}^{n-1} p_i^2 + q_i^2 < \delta \right\},$$

the following relations take place

$$\varphi_1(p, q) \geq 0, \quad \varphi_2(p, q) \geq 0, \quad \varphi_1(p, q) + \varphi_2(p, q) > 0,$$

and

$$B(p) = B \cap R^n(p) < \omega(p), \quad 0 < p_i < \tilde{\varepsilon}.$$

That is why, together with inequalities 1–4, at least one of equalities (1.16) cannot be performed for all $p \in \mathbf{R}^n$. The theorem is proved.

1.6 EXISTENCE AND UNIQUENESS OF A SOLUTION TO AN INVERSE PROBLEM FOR A PARABOLIC EQUATION

Problems of finding the coefficients of differential equations from information about their solutions are called inverse problems for differential equations. As a rule, inverse problems are nonlinear. Their study, especially in the multidimensional case, is often connected with significant mathematical difficulties. We will give a method for studying solvability questions for certain nonlinear inverse problems for differential equations, using a parabolic equation as an example illustrating this method.

With the aid of the Fourier transform we are able to reduce the inverse problem under consideration to a boundary value problem for a nonlinear integro-differential equation which is fully acceptable for investigation. Methods of potential theory are applicable to the boundary value problem thus obtained; under the appropriate restrictions on the data, this leads to determination of solvability, uniqueness, and stability conditions for this inverse problem.

In a domain $Q = (-\infty, \infty) \times D$, $D \subset \mathbf{R}^n$, $\partial D = \Gamma_0 \in C$, we consider the equation of parabolic type

$$\rho(x)u_t - \Delta u = 0. \tag{1.17}$$

Inverse problem. Find function $u(x,t)$, $\rho(x) > 0$ satisfying (1.17) such that

$$u|_\Gamma = \varphi(s,t), \quad \Gamma = \Gamma_0 \times (-\infty, \infty), \tag{1.18}$$

$$u\big|_{t=0} = u_0(x), \quad x \in D, \quad u_0(x)\big|_{x=s} = \varphi|_{t=0}\,, \tag{1.19}$$

$$\nabla_x u \in L^2(Q), \quad u \in L^2(-\infty, \infty : W_2^2(D) \cap C(\bar{D})), \tag{1.20}$$

$$u_t \in L^\infty(-\infty, \infty) : C(\bar{D}), \quad \rho(x) \in C(\bar{D}).$$

We introduce some notation. $G(x,y)$ is Green's function of the Dirichlet problem for the domain D, $\hat{u}(x,\xi)$ is the Fourier transform of a function $u(x,t)$ with respect to t and

$$\mu(D) = \max_{\bar{D}} \int_D |G(x,y)|\, dy.$$

First we consider problem (1.17) and (1.18).

Lemma 1.3. *Let $\hat{\varphi}(\xi, s) \in W_2^{s/2}(\Gamma_0) \cap C(\Gamma_0)$; $\hat{\varphi} \equiv 0$, $|\xi| > R$, and $s \in \Gamma_0$. Then, for any real-valued function $\rho(x)$ such that*

$$\rho(x) \in C(\bar{D}), \quad |\rho| \le 1/R\mu(D),$$

there is a unique solution to problems (1.17) and (1.18) such that $\nabla_x u \in L^2(Q)$, $u \in L^2(-\infty, \infty : W_2^2(D) \cap C(\bar{D}))$, $u_t \in L^\infty((-\infty,\infty) : C(\bar{D}))$ and $\hat{u}(x,\xi) \equiv 0$, $|\xi| > R$, $x \in \bar{D}$.

A proof of Lemma 1.3 is based on the study of the boundary value problem

$$\Delta v + i\xi\rho(x)v = 0, \quad v|_{\Gamma_0} = \hat{\varphi}(x,\xi), \quad x \in D,$$

and is carried out as in (Bubnov, 1984).

EXAMPLE. Let $A(\xi)$ and $B(\xi)$ be continuous functions, $0 \leq \xi \leq R_0$, $R_0 > 0$, let ρ be a positive constant, and let $u(x,t)$, $x,t \in \mathbf{R}^1$ be the function

$$u(x,t) = \int_0^{R_0} \exp\left(x\sqrt{\frac{\xi}{2p}}\right)\left[\cos\left(x\sqrt{\frac{\xi}{2p}} + \xi t\right) A(\xi) + B(\xi)\sin\left(x\sqrt{\frac{\xi}{2p}} + \xi t\right)\right] \mathrm{d}\xi,$$

that is entire analytic in t and satisfies the heat equation

$$\frac{\partial u}{\partial t} = \rho\frac{\partial^2 u}{\partial x^2}.$$

According to the Paley-Wiener theorem,

$$\hat{u}(x,\xi) \equiv 0, \quad |x| > R_0.$$

Our study of the inverse problem is based on the following lemmas. Let

$$A_1 = \sup_{x,\xi}\left|\int_{\Gamma_0} \hat{\varphi}\frac{\partial G}{\partial n}\mathrm{d}\Gamma_0\right|,$$

$$A_2 = \min_{\bar{D}}\left|-\int_{|\xi|<R} \mathrm{i}\xi \int_{\Gamma_0} \hat{\varphi}\frac{\partial G}{\partial n}\mathrm{d}\Gamma_0 \mathrm{d}\xi\right|, \quad \omega_0 = \mu(D)\left\|\Delta u_0\right\|_C R.$$

Lemma 1.4. *Let the conditions of Lemma 1.3 hold, and let*

$$A_2 = \min_{\bar{D}}\left[-\int_{|\xi|<R} \mathrm{i}\xi \int_{\Gamma_0} \hat{\varphi}\frac{\partial G}{\partial n}\mathrm{d}\Gamma_0 \mathrm{d}\xi\right] > 0,$$

$$A_2 \geq \omega_0 + R(8\omega_0 A_1)^{1/2}, \quad 8R^2 A_1 > \omega_0.$$

Then the inverse problem (1.17)–(1.20) is equivalent to the boundary value problem for the nonlinear integro-differential equation

$$\Delta_x v(x,\xi) + \frac{\mathrm{i}\xi\Delta u_0(x,\xi)}{-\int_{|\xi|<R} \mathrm{i}\xi v(x,\xi)\mathrm{d}\xi} = 0, \quad v|_{\Gamma_0} = v_0(s,\xi) \tag{1.21}$$

in the class of functions $v \in C(\bar{D}) \cap W_2^2(D)$, $v(x,\xi) \equiv 0$, $|x| > R$, $x \in \bar{D}$;

$$\min_{\bar{D}}\left[-\int_{|\xi|<R} \mathrm{i}\xi v(x,\xi)\mathrm{d}x\right] \geq R_0 > 0,$$

$$R_0 = \frac{1}{3}\left[\omega_0 + A_2 + \left[(\omega_0 - A_2)^2 - 8R^2\omega_0 A_1\right]^{1/2}\right].$$

Taking the nonlinearity into account and using methods of potential theory, we prove the following result.

Lemma 1.5. *Let the conditions of Lemma 1.4 be satisfied. Then there is $\delta(R, A_1, A_2) > 0$ such that for any u_0 with $0 < \Delta u_0 < \delta$ there exists a unique solution $v(x,\xi)$ to problem (1.21) that belongs to the class*

$$B = \Big\{ v(x,\xi) \in C(\bar{D}) \cap W_2^2(D), \quad v(x,\xi) \equiv 0, \quad |\xi| > R, \quad x \in \bar{D},$$

$$\sup_{x,\xi} |v(x,\xi)| \le \frac{R_0 A_1}{R_0 - \omega_0}, \quad \min_{\bar{D}} \Big[- \int_{|\xi|<R} i\xi v(x,\xi) d\xi \Big] \ge R_0 \Big\}.$$

From the lemmas formulated above we get our main result.

Theorem 1.10. *Let the following conditions be satisfied:*

$$\hat{\varphi}(\xi, s) \in W_2^{s/2}(\Gamma_0) \cap C(\Gamma_0), \quad \hat{\varphi} \equiv 0, \quad |\xi| > R,$$

$$\hat{\varphi}|_{t=0} = u_0(x)|_0, \quad 0 < \Delta u_0 \in C(\bar{D}),$$

$$\min_{\bar{D}} \Big[- \int_{|\xi|<R} i\xi \int_{\Gamma_0} \hat{\varphi} \frac{\partial G}{\partial n} d\Gamma_0 d\xi \Big] = A_2 > 0,$$

$$A_2 \ge \omega_0 + R\,(8\omega_0 A_1)^{1/2}, \quad 8R^2 A_1 > \omega_0.$$

Then there is a $\delta(R, A_1, A_2) > 0$ such that for any u_0 with $0 < u_0 \le \delta$ the inverse problem (1.17)–(1.20) has a unique solution(ρ, u); moreover,

$$\rho(x) = \frac{\Delta u_0(x,\xi)}{- \int_{|\xi|<R} i v(x,\xi) d\xi},$$

$$u(x,t) = \int_{-\infty}^{\infty} v(x,\xi) \exp(-\xi t) dx,$$

where $v(x,\xi)$ is a solution to problem (1.21).

REMARK T. his theorem can be briefly formulated as follows: if in addition to the conditions of the first boundary value problem for a parabolic equation we require that a solution be regular in t, $-\infty < t < \infty$, then, under suitable restrictions on the data of the boundary value problem, one can also uniquely find the function $\rho(x)$ occurring in the equation, i.e. one can solve the inverse problem. From the method it is clear that this fact permits a significant generalization for other boundary value problems and other evolution equations including equation with variable coefficients.

1.7 ON UNIQUE SOLVABILITY OF AN INVERSE PROBLEM FOR A PARABOLIC EQUATION

We formulate conditions for the unique solvability of the Cauchy problem for a semilinear integro-differential equation, and offer a method for its investigation (Anikonov and Belov, 1989). Such problems are reductions of inverse problems for parabolic equations in cases where the Fourier transform of their solutions with respect to chosen variables exists and some other conditions are satisfied (see (Anikonov, 1986)). Analogous inverse problems for parabolic and hyperbolic equations were investigated by different methods in (Bubnov, 1987a,b).

Let $(x_1, \dots, x_{n-1})$ be a point in the Euclidean space $\mathbf{R}^{n-1}$,

$\Pi_{[t_1,t_2]} = \{(t,x) \mid t_1 \le t \le t_2, \quad x \in \mathbf{R}^{n-1}\}$ a strip in $\mathbf{R}^n$,

$G_{[t_1,t_2]} = \left\{(t,x,z) \mid (t,x) \in \Pi_{[t_1,t_2]}, \quad z \in \mathbf{R}^1\right\}$ a strip in $\mathbf{R}^n$,

$G^\alpha = \{(x,z) \mid x \in \mathbf{R}^{n-1}, \quad z \in [-\alpha,\alpha]\}$,

$G^\alpha_{[t_1,t_2]} = \left\{(t,x,z) \mid (t,x) \in \Pi_{[t_1,t_2]}, \quad z \in [-\alpha,\alpha]\right\}$ and $\alpha = \text{const} > 0$.

Let us examine the following equation in $G_{[0,T]}$:

$$\frac{\partial u(t,x,z)}{\partial t} = L(u(t,x,z)) + \frac{\partial^2 u(t,x,z)}{\partial t^2} + a(t,x)\frac{\partial u(t,x,z)}{\partial z} + q(t,x,z), \tag{1.22}$$

where

$$L(u) = \sum_{i,j=1}^{n-1} a_{ij}\frac{\partial^2 u}{\partial x_i \partial x_j} + \sum_{i=1}^{n-1} b_i \frac{\partial u}{\partial x_i} + cu.$$

We assume that a_{ij}, b_i and c are sufficiently smooth functions of t given on $[0,T]$ and, in addition, $c(t) \le 0$ and

$$\mu\,|\xi|^2 \le \sum_{i,j=1}^{n-1} a_{ij}(t)\xi_i\xi_j, \quad \forall \xi \in \mathbf{R}^{n-1}, \quad t \in [0,T], \quad \mu = \text{const} > 0. \tag{1.23}$$

The function $q(t,x,z)$ is given on $G_{[0,T]}$.

The function $a(t,x)$ $(\partial a/\partial z \equiv 0)$ has yet to be found. We assume that the following condition is satisfied:

$$u\big|_{z=0} = \varphi(t,x), \quad (t,x) \in \Pi_{[0,T]}, \tag{1.24}$$

and that the Fourier transform of $u(t,x,z)$ with respect to z exists:

$$\begin{aligned} W(t,x,y) &= \frac{1}{2\pi}\int_{-\infty}^{\infty} u(t,x,y)\,\mathrm{e}^{-izy}\mathrm{d}z, \\ u(t,x,y) &= \int_{-\infty}^{\infty} W(t,x,y)\,\mathrm{e}^{izy}\mathrm{d}y. \end{aligned} \tag{1.25}$$

Let us introduce a Cauchy condition for equation (1.22):

$$u(0,x,z) = u_0(x,z), \quad (x,z) \in \mathbf{R}^n. \tag{1.26}$$

With the help of conditions (1.24) and (1.25), problem (1.22) and (1.26) can be reduced to a Cauchy problem for a integro-differential equation, not containing $a(t,x)$. Indeed, by applying the Fourier transform to equation (1.22), setting $z=0$ in (1.22) under conditions (1.24) and (1.25), we obtain the equations

$$\frac{\partial W}{\partial t} = L(W) - y^2 W + \mathrm{i}yaW + Q,$$

$$\frac{\partial \varphi}{\partial t} = L(\varphi) - \int_{-\infty}^{\infty} \lambda^2 W \mathrm{d}\lambda + \mathrm{i}a \int_{-\infty}^{\infty} \lambda W \mathrm{d}\lambda + q(t,x,0),$$

from which we eliminate function a and obtain the problem

$$\begin{gathered} \frac{\partial W}{\partial t} = L(W) - y^2 W + \frac{yM}{\int_{-\infty}^{\infty} \lambda W \mathrm{d}\lambda} \left\{ \Phi + \int_{-\infty}^{\infty} \lambda^2 W \mathrm{d}\lambda \right\} + Q, \\ W(0,x,y) = W_0(x,y), \quad (x,y) \in \mathbf{R}^n. \end{gathered} \tag{1.27}$$

Here

$$Q(t,x,y) = \frac{1}{2\pi} \int_{-\infty}^{\infty} q(t,x,y)\,\mathrm{e}^{-\mathrm{i}zy} \mathrm{d}z$$

is the Fourier transform of q and

$$W_0(x,y) = \frac{1}{2\pi} \int_{-\infty}^{\infty} u_0(x,y)\,\mathrm{e}^{-\mathrm{i}zy} \mathrm{d}z,$$

$$\Phi(t,x) = \frac{\partial \varphi(t,x)}{\partial t} - L(\varphi(t,x)) - q(t,x,0).$$

We examine the simplest case, when Q and W_0 are real-valued functions with compact supports in y, belonging to a finite segment $[-\alpha,\alpha]$. Then (1.27) reduces to the problem

$$\frac{\partial W}{\partial t} = L(W) - y^2 W + \frac{yW}{\int_{-\alpha}^{\alpha} \lambda W \mathrm{d}\lambda} \left\{ \Phi + \int_{-\alpha}^{\alpha} \lambda^2 W \mathrm{d}\lambda \right\} + Q, \tag{1.28}$$

$$W(0,x,y) = W_0(x,y), \quad x \in \mathbf{R}^{n-1}, \quad y \in [-\alpha,\alpha]. \tag{1.29}$$

Equation (1.28) is a nonlinear integro-differential. It consists of partial derivatives as well as integrals of the solution with respect to the parameter, which also appears in the coefficients and the initial data of the problem. These are interesting mathematical objects which have not yet been studied. We examine the solvability of problems (1.28) and (1.29) and some properties of its solutions.

Below we denote by $C^{k,l}(G^d_{[t_1,t_2]})$ the space of functions that have continuous derivatives in $G^\alpha_{[t_1,t_2]}$ in the space variables of order up to and including k, and a derivative of order l in the variable t.

As to the input data of problem (1.28) and (1.29), we presume the following:

$$\begin{cases} \Phi, q, W_0 \in (G^\alpha_{[0,T]}); \text{ all } k\text{th derivatives in } x \text{ and } y \\ \qquad\qquad \text{are bounded in absolute value} \\ \text{by constants } c_k, \quad k = 0,\dots,4, \text{ respectively in } G^\alpha_{[0,T]} \end{cases} \tag{1.30}$$

$$Q(t,x,y)y \geq 0, \quad W_0(x,y)y \geq 0, \quad t \in [0.T], \quad x \in \mathbf{R}^{n-1}, \quad y \in [-\alpha,\alpha]; \tag{1.31}$$

$$I_{W_0}(x) \equiv \int_{-\alpha}^{\alpha} \lambda W_0(x,\lambda)\mathrm{d}\lambda \geq \delta > 0, \quad x \in \mathbf{R}^{n-1}. \tag{1.32}$$

Let us split problem (1.28) and (1.29) according to the method of the weak approximation (Yanenko, 1971; Belov, 1986):

$$\frac{1}{2}\frac{\partial W_\tau}{\partial t} = L(W_\tau), \quad n\tau < t \leq (n+\frac{1}{2})\tau, \tag{1.33}$$

$$\frac{1}{2}\frac{\partial W_\tau}{\partial t} = -y^2 W_\tau + \frac{yW_\tau}{\int_{-\alpha}^{\alpha} \lambda W_\tau \mathrm{d}\lambda}\left\{\Phi + \int_{-\alpha}^{\alpha} \lambda^2 W_\tau \mathrm{d}\lambda\right\} + Q, \quad (n+\frac{1}{2})\tau < t \leq (n+1)\tau, \tag{1.34}$$

$$W_\tau(0,x,y) = W_0(x,y), \quad (x,y) \in G^\alpha. \tag{1.35}$$

Here $n = 0,\ldots,N-1$ and $N_\tau = t_*$, $0 < t_* \leq T$.

We start with the Cauchy problem

$$\frac{1}{2}\frac{\partial v(t,x,y)}{\partial t} = \frac{yv(t,x,y)}{\int_{-\alpha}^{\alpha} \lambda v(t,x,y)\mathrm{d}\lambda}\left\{\Phi + \int_{-\alpha}^{\alpha} \lambda^2 v(t,x,y)\mathrm{d}\lambda\right\} - y^2 v(t,x,y) + Q(t,x,y), \tag{1.36}$$

$$v(t_0,x,y) = v_0(x,y), \quad (x,y) \in G^\alpha. \tag{1.37}$$

By (1.30'), (1.31') and (1.32') we denote conditions (1.30), (1.31), and (1.32) with W_0 replaced by v_0.

Lemma 1.6. *Let $\Phi \leq k$, $k = \mathrm{const}$, and let conditions (1.30')- (1.32') be satisfied. Then there exists ν_0, $0 < \nu_0 \leq T$, such that problem (1.36) and (1.37) is uniquely solvable in the class $C^{4,1}(G^\alpha_{[t_0,t_0+\nu]})$ for any $\nu \in (0,\nu_0]$ and any $t_0 \in [0,T-\nu]$. ν_0 depends only on the constants δ, k and α.*

PROOF. Let us linearize problem (1.36) and (1.37) by making a shift $\theta > 0$ with respect to t of some terms in (1.36):

$$\begin{aligned}\frac{1}{2}\frac{\partial v^\theta(t,x,y)}{\partial t} &= \frac{yv^\theta(t,x,y)}{\int_{-\alpha}^{\alpha} \lambda v^\theta(t-\theta,x,y)\mathrm{d}\lambda}\left\{\Phi + \int_{-\alpha}^{\alpha} \lambda^2 v^\theta(t-\theta,x,y)\mathrm{d}\lambda\right\} \\ &- y^2 v^\theta(t,x,y) + Q(t,x,y),\end{aligned} \tag{1.38}$$

$$v^\theta\Big|_{t\leq t_0} = v_0(x,y), \quad (x,y) \in G^\alpha.$$

It can be shown that there exists ν_0 depending on δ, k, and α such that for all $\nu \in (0,\nu_0]$ and $t_0 \in [0,T-\nu]$ the solution v^θ of problem (1.38) exists in the interval $[t_0,t_0+\nu]$ for any fixed $\theta \in (0,\tilde{\theta}]$ and uniformly in $\theta \in (0,\tilde{\theta}]$ and $(t,x) \in \Pi_{[t_0,t_0+\nu]}$.

$$\int_{-\alpha}^{\alpha} \lambda v^\theta(t-\theta,x,y)\mathrm{d}\lambda \geq \mu > 0. \tag{1.39}$$

Estimate (1.39) allows us to obtain uniform boundedness (in θ) of the family of solutions $\{v^\theta\}$ and their first derivatives with respect to x and y. Hence, as $\theta \to 0$, uniformly in every compactum of $G^{\alpha}_{[t_0,t_0+\nu]}$, the functions v^{θ} converge to v being solution of the problem (1.36) and (1.37) in $C^{4,1}(G^{\alpha}_{[t_0,t_0+\nu]})$. As a result

$$\int_{-\alpha}^{\alpha} \lambda v(t-\theta,x,y)\mathrm{d}\lambda \geq \mu > 0, \quad (t,x) \in \Pi_{[t_0,t_0+\nu]}.$$

This allows one to prove the uniqueness of the solution in the class specified above. The lemma is proved.

REMARK 1.1. In consequence of condition (1.31'), the inequality $v(t,x,y)y \geq 0$, $(t,x,y) \in G^{\alpha}_{[t_0,t_0+\nu]}$ is realized.

REMARK 1.2. Let v be the solution of problems (1.36), (1.37) in the interval $[t_0, t_0+\nu]$, the existence of which is guaranteed by Lemma 1.6. As follows from remark 1.1, we can apply the generalized mean value theorem (Fikhtengol'tz, 1966) to the integrals $\int_{-\alpha}^{\alpha} \lambda^2 v\mathrm{d}\lambda$ and $\int_{-\alpha}^{\alpha} \lambda^3 v\mathrm{d}\lambda$:

$$\int_{-\alpha}^{\alpha} \lambda^{k+1} v(t,x,\lambda)\mathrm{d}\lambda = m_k(t,x) \int_{-\alpha}^{\alpha} \lambda v(t,x,\lambda)\mathrm{d}\lambda, \quad k=1,2. \tag{1.40}$$

It is clear that m_k are smooth functions that satisfy $-\alpha < m_1(t,x) < \alpha$ and $0 < m_2(t,x) < \alpha^2$. By multiplying (1.36) by y, integrating the result over $[-\alpha,\alpha]$ and then applying (1.40), we find that

$$I_V(t,x) \equiv \int_{-\alpha}^{\alpha} \lambda v(t,x,\lambda)\mathrm{d}\lambda$$

is a solution of an ordinary differential equation

$$\frac{\mathrm{d}I_v(t,x)}{\mathrm{d}t} = m_1(t,x)\Phi(t,x) + (m_1^2(t,x) - m_2(t,x))I_v(t,x) + I_Q,$$

where x is a parameter. Let us consider the equation with constant coefficients

$$\frac{\mathrm{d}j(t)}{\mathrm{d}t} = -k\alpha - \alpha^2 j(t). \tag{1.41}$$

Its solution $j(t)$, satisfying the initial condition $j(0) = \delta$, is monotonic decreasing and strictly positive in an interval $[0,t_*]$. The value t_* depends on δ, k and α. Since $m_1\Phi \geq -\alpha k$ and $m_1^2 - m_2 > -\alpha^2$ we have

$$I_V(t,x) \geq j(t) \geq j(t_*), \quad t_0 \leq t \leq t_*, \tag{1.42}$$

provided that

$$I_v(t_0,x) \equiv \int_{-\alpha}^{\alpha} \lambda v_0(x,\lambda)\mathrm{d}\lambda \geq j(t_0).$$

Theorem 1.11. *Let $|\Phi| \leq k =$ const, and let conditions (1.30)–(1.32) be satisfied. Then there exists $t_* \in (0,T]$ such that problem (1.28) and (1.29) are uniquely solvable in $C^{2,1}(G^{\alpha}_{[0,t_*]})$. The value of t_* depends on the constants δ, k, and α.*

PROOF. By applying the maximum principle to equation (1.33) (see conditions (1.22), Lemma 1.6, and the inequalities (1.42) of remark 1.2), we can easily show that a solution $W_\tau(t,x,y)$ of problems (1.33)–(1.35) exists in the interval $[0,t_*]$ for any fixed $\tau \in (0,\tau_0]$ and, uniformly in $\tau \in (0,\tau_0]$,

$$I_{W_\tau}(t,x) \geq j(t) \geq j(t_*), \quad t_0 \leq t \leq t_*, \quad (t,x) \in \Pi_{[0,t_*]}, \tag{1.43}$$

where $j(t)$ is a solution of (1.41) and $j(0) = \delta$. Estimate (1.43), the maximum principle for equation (1.33), and condition (1.30) allow us to prove the estimates, uniform in $\tau \in (0,\tau_0]$,

$$\sum_{|k|\leq 4} \sup_{G^\alpha_{[0,t_*]}} \left|\mathbf{D}^k W_\tau\right| + \sum_{|k|\leq 2} \sup_{G^\alpha_{[0,t_*]}} \left|\mathbf{D}_t \mathbf{D}^k W_\tau\right| \leq c, \tag{1.44}$$

where $\mathbf{D}_t = \partial/\partial t$, $\mathbf{D}^k = \partial^{|k|}/\partial x_1^{k_1} \ldots x_n^{k_{n-1}} \partial y^{k_n}$, $k = (k_1, \ldots, k_n)$ is a multi-index, and $|k| = k_1 + \ldots + k_n$. Because of (1.44) as in (Yanenko, 1971) and (Belov, 1986), we can prove that, uniformly for each compactum of $G^\alpha_{[0,t_*]}$, W_τ converges to $W \in C^{2,1}(G^\alpha_{[0,t_*]})$ along with its derivatives of first and second order with respect to the space variables x, y, and is a solution of problem (1.28) and (1.29). The uniqueness of the solution is proved in a standard way by proving that the difference of two possible solutions of problem (1.28) and (1.29) from the class $C^{2,1}(G^\alpha_{[0,t_*]})$ is identically zero. The theorem is proved.

REMARK 1.3. If $\Phi = 0$, then we can take $k = 0$, and in this case the solution $j(t)$ of equation (1.41) with the initial condition $j(0) = \delta$ is strictly positive in the whole interval $[0,T]$. This guarantees the solvability of problem (1.28) and (1.29) in this interval.

1.8 FORMULAS IN MULTIDIMENSIONAL INVERSE PROBLEMS FOR EVOLUTION EQUATIONS

Let D be a domain in the real Euclidean space $\mathbf{R}^{n+m+1}$ of variables (x,z,t), $x = (x_1, \ldots, x_m)$, $z = (z_1, \ldots, z_n)$, $t \in \mathbf{R}^1$. We consider the inverse problem of the simultaneous determination of two complex-valued functions $u(x,z,t)$ and $\lambda(x,t)$ ($\partial\lambda/\partial z_k = 0$, $i = 1, \ldots, n$) in D which satisfy

$$\frac{\partial u}{\partial t} = \mathbf{A}u + \mathbf{L}u + \lambda(x,t)\mathbf{B}u. \tag{1.45}$$

$$u\big|_{t=0} = u_0(x,z), \quad u\big|_{z=0} = f(x,t). \tag{1.46}$$

Here $\mathbf{A}$, $\mathbf{B}$, and $\mathbf{L}$ are linear differential operators such that $\mathbf{A}$ and $\mathbf{B}$ act with respect to x and are independent of (z,t) while $\mathbf{L}$ acts with respect to z and is independent of (x,t), the functions $u_0(x,z)$ and $f(x,t)$ are given, and $(x,z,t) \in D$.

As examples of operators $\mathbf{A}$, $\mathbf{B}$, and $\mathbf{L}$ we may consider linear differential operator with constant coefficients $\mathbf{A}_\alpha$, $\mathbf{B}_\alpha$, and $\mathbf{L}_\alpha$ such that

$$\mathbf{A} = \sum_\alpha \mathbf{A}_\alpha \mathbf{D}^\alpha_x, \quad \mathbf{B} = \sum_\alpha \mathbf{B}_\alpha \mathbf{D}^\alpha_x, \quad \mathbf{L} = \sum_\alpha \mathbf{L}_\alpha \mathbf{D}^\alpha_z.$$

In particular, with

$$\mathbf{A} = \sum_{j=1}^m \frac{\partial^2}{\partial x_j^2}, \quad \mathbf{L} = \frac{\partial^2}{\partial z_1^2}, \quad n = 1,$$

arises the inverse problem for parabolic equation, namely, to find functions $u(x,z,t)$ and $\lambda(x,t)$ satisfying

$$\frac{\partial u}{\partial t} = \Delta u + \lambda(x,t)\mathbf{B}u, \quad u|_{t=0} = u_0(x,z), \quad u|_{z=0} = f(x,t).$$

For $\mathbf{B}u = \varphi(z,t)$, where $\varphi(z,t)$ is some function, the inverse problem (1.45) and (1.46) consists in finding a source function J of the form $J = \lambda(x,t)\varphi(z,t)$.

We study the case when the initial state $u|_{t=0} = u_0(x,z)$ can be factored in the form of a product $a(x)b(z)$, and establish explicit formulas for the solutions $u(x,z,t)$, $\lambda(x,t)$ of (1.45) and (1.46), which introduces a constructive element in the theory of such problems. In what follows we assume that **A**, **B**, and **L** act in a well-defined way on the functions under consideration, and the integral and other transformations used are also well defined.

Lemma 1.7. *Suppose that the function $f(x,t)$ such that $\mathbf{B}f \not\equiv 0$, that $\varphi(z,t)$ is a solution of the equation*

$$\frac{\partial \varphi}{\partial t} = \mathbf{L}\varphi,$$

and that $\varphi_0(t) = \varphi(0,t)$, $(x,z,t) \in D$. Then the functions

$$u(x,z,t) = \frac{f(x,t)\varphi(z,t)}{\varphi_0(t)}, \quad \lambda(x,t) = \frac{\left(\dfrac{\partial f}{\partial t} - \mathbf{A}f\right)\varphi_0 - f\dfrac{\mathrm{d}\varphi_0}{\mathrm{d}t}}{Bf\varphi_0}$$

satisfy equation (1.45) in D and the conditions

$$u|_{z=0} = f(x,t), \quad u|_{t=0} = a(x)b(z),$$

where $a(x) = f(x,0)$ and $b(z) = \varphi(z,0)/\varphi_0(0)$.

Let **L** be a differential operator with constant coefficients and $c(\omega)$ its symbol; that is $\mathbf{L}(\mathrm{e}^{\mathrm{i}\omega z}) = c(\omega)\,\mathrm{e}^{\mathrm{i}\omega z}$, $\omega \in \mathbf{R}^n$.

Lemma 1.8. *Let $f(x,t)$ and $q(\omega)$, $(x,t) \in D$, $\omega \in \mathbf{R}^n$ be complex- valued functions such that the expressions*

$$\frac{\partial f}{\partial t} = \mathbf{A}f, \quad \mathbf{B}f, \quad \int\limits_{\mathbf{R}^n} \mathrm{e}^{c(\omega)t+\mathrm{i}\omega z} q(\omega)\mathrm{d}\omega$$

are meaningful and $\mathbf{B}f \not\equiv 0$, $\int\limits_{\mathbf{R}^n} \mathrm{e}^{c(\omega)t} q(\omega)\mathrm{d}\omega \neq 0$. Then the functions

$$u(x,z,t) = \frac{f(x,t)\displaystyle\int\limits_{\mathbf{R}^n} \mathrm{e}^{c(\omega)t+\mathrm{i}\omega z} q(\omega)d\omega}{\displaystyle\int\limits_{\mathbf{R}^n} \mathrm{e}^{c(\omega)t} q(\omega)\mathrm{d}\omega},$$

$$\lambda(x,t) = \frac{\left(\dfrac{\partial f}{\partial t} - \mathbf{A}f\right)\displaystyle\int\limits_{\mathbf{R}^n} \mathrm{e}^{c(\omega)t} q(\omega)\mathrm{d}\omega - f\displaystyle\int\limits_{\mathbf{R}^n} \mathrm{e}^{c(\omega)t} q(\omega)c(\omega)\mathrm{d}\omega}{\mathbf{B}f\displaystyle\int\limits_{\mathbf{R}^n} \mathrm{e}^{c(\omega)t} q(\omega)\mathrm{d}\omega}$$

satisfy equation (1.45) and the conditions $u|_{z=0} = f(x,t)$ *and* $u|_{t=0} = a(x)b(z)$, *where*

$$a(x) = f(x,0), \quad b(z) = \frac{\int\limits_{\mathbf{R}^n} q(\omega)\, e^{i\omega z} d\omega}{\int\limits_{\mathbf{R}^n} q(\omega) d\omega}.$$

The following assertion holds.

Theorem 1.12. *Suppose that* $\mathbf{L} = \sum_1^n \frac{\partial^2}{\partial z_k^2}$, $z \in \mathbf{R}^n$, *and* $u_0(x,z) = a(x)b(z)$, *in problem (1.45) and (1.46), where* $a(x) = b(x,0)$ *and* $b(z)$ *is a continuous bounded complex-valued function in* $\mathbf{R}^n$ *with* $b(0) = 1$. *If the conditions in Lemma 1.7 hold, then the solution* $u(x,z,t)$, $\lambda(x,t)$ *of the inverse problem (1.45) and (1.46) can be represented in the form*

$$u(x,z,t) = \frac{f(x,t) \int\limits_{\mathbf{R}^n} b(y) \exp(-|y-z|^2/4t) dy}{\int\limits_{\mathbf{R}^n} b(y) \exp(-y^2/4t) dy},$$

$$\begin{aligned} \lambda(x,t) \;=\; & \left\{ \left(\frac{\partial f}{\partial t} - \mathbf{A} f \right) \frac{1}{(2\pi t)^{n/2}} \int\limits_{\mathbf{R}^n} b(y) \exp\left(-\frac{y^2}{4t}\right) dy \right. \\ & - \left. f \frac{\partial}{\partial t} \frac{1}{(2\pi t)^{n/2}} \int\limits_{\mathbf{R}^n} b(y) \exp\left(-\frac{y^2}{4t}\right) dy \right\} \\ & \times \left(\mathbf{B} f \frac{1}{(2\pi t)^{n/2}} \int\limits_{\mathbf{R}^n} b(y) \exp\left(-\frac{y^2}{4t}\right) dy \right)^{-1}. \end{aligned}$$

COROLLARY . *Suppose that* $\mathbf{B}u = \varphi(z,t)$ *in (1.45) and that the conditions of the Theorem 1.12 hold. Then the source function* $J(x,z,t) = \varphi(z,t)\lambda(x,t)$ *and the solution* $u(x,z,t)$ *of (1.45) can be found from the data (1.46) by means of the formulas*

$$\varphi(z,t) = \frac{1}{(2\pi t)^{n/2}} \int\limits_{\mathbf{R}^n} b(y) \exp(-|y-z|^2/4t) dy,$$

$$u(x,z,t) = \frac{f(x,t)\varphi(z,t)}{\varphi_0(t)}, \quad \varphi_0(t) = \varphi(0,t),$$

$$J(x,z,t) = \frac{\left(\frac{\partial f}{\partial t} - \mathbf{A} f \right) \varphi_0 - f \frac{d\varphi_0}{dt}}{\varphi_0^2(t)} \varphi(z,t).$$

From Lemma 1.8 we deduce the following assertion.

Theorem 1.13. *Suppose that* $u_0(x,z) = a(x)b(z)$ *in (1.45) and (1.46), where* $z \in \mathbf{R}^n$, $a(x) = f(x,0)$, $b(z)$ *has a Fourie transform and* $b(0) = 1$, *and that the conditions in*

Lemma 1.8 hold. Then the solution $u(x,z,t)$, $\lambda(x,t)$ of the inverse problem (1.45) and (1.46) is given by

$$u(x,z,t) = \frac{f(x,t) \int\limits_{\mathbf{R}^n} \mathrm{e}^{c(\omega)t + \mathrm{i}\omega z} q(\omega) \mathrm{d}\omega}{\int\limits_{\mathbf{R}^n} \mathrm{e}^{c(\omega)t} q(\omega) \mathrm{d}\omega},$$

$$\lambda(x,t) = \frac{\left(\frac{\partial f}{\partial t} - \mathbf{A} f\right) \int\limits_{\mathbf{R}^n} \mathrm{e}^{c(\omega)t} q(\omega) \mathrm{d}\omega - f \int\limits_{\mathbf{R}^n} \mathrm{e}^{c(\omega)t} q(\omega) c(\omega) \mathrm{d}\omega}{\mathbf{B} f \int\limits_{\mathbf{R}^n} \mathrm{e}^{c(\omega)t} q(\omega) \mathrm{d}\omega},$$

where

$$q(\omega) = \int\limits_{\mathbf{R}^n} b(y)\, \mathrm{e}^{-\mathrm{i}z\omega} \mathrm{d}z.$$

CHAPTER 2

Inverse Problems for Kinetic Equations

2.1 KINETIC EQUATIONS

Kinetic equations characterize the continuous nature of motion of an object, they are the fundamental equations of the natural sciences. These equations are widely used for qualitatively describing physical, chemical, biological and other processes at the microscopic level. In view of the great role kinetic equations have, as a rule, integro-differential equations can be used. These are often nonlinear relative to one or several distribution functions $F(x,p,t)$ describing the dynamic state of an object with respect to the spatial position x, the momentum p and the time t. The Liouville, Boltzmann, Vlasov equations, a chain of the Bogolyubov equations, the Fokker-Planck equations, various equations of radiation transport, quantum equations for the density matrix and the Wigner function (see Alekseev, 1983; Bogolyubov, 1981; Case and Zweifel, 1967; Ferzeiger and Kaper, 1972; Haken, 1987; Klimontovich, 1982; Kompaneets, 1977; Lifshitz and Pitaevskij, 1979; Prigogine, 1980) are examples of kinetic equations.

Below $F(x,p,t)$ is a distribution function.

1. Liouville equation:

$$\frac{\partial F}{\partial t} + \{F, H\} = 0, \quad \{F, H\} = \sum_{j=1}^{n} \left(\frac{\partial F}{\partial x_j}\frac{\partial H}{\partial p_j} - \frac{\partial F}{\partial p_j}\frac{\partial H}{\partial x_j} \right).$$

2. Boltzmann equation:

$$\frac{\partial F}{\partial t} + \sum_{j=1}^{n} \frac{\partial F}{\partial x_j} p_j = \mathrm{St}\, F.$$

3. Transport equation:

$$\frac{\partial F}{\partial t} + \Omega\, \mathrm{grad}_x\, F + \sigma F = \int\limits_{\omega} I(x,t,\Omega,\Omega') F \mathrm{d}\Omega' + q.$$

4. Transport equation in tomography:

$$\frac{\partial F}{\partial t} + \Omega\, \mathrm{grad}_x\, F + \sigma F = \int\limits_{E_1}^{E_2} \int\limits_{\omega} I(x,t,E\prime,\Omega,\Omega') F \mathrm{d}E' \mathrm{d}\Omega' + q.$$

5. Kinetic equation in problems of integral geometry:

$$\Omega \operatorname{grad}_x F\Omega + \sigma F = 0, \quad \ln F = \psi \to \operatorname{grad}_x \psi\Omega = -\sigma.$$

$$\frac{\partial F}{\partial x}\cos\varphi + \frac{\partial F}{\partial y}\sin\varphi + K(x,y,\varphi)\frac{\partial F}{\partial \varphi} = \lambda(x,y,\varphi),$$

K - first curvature.

6. Fokker-Planck equation:

$$\frac{\partial F}{\partial t} + \{F,H\} = \int_{\mathbf{R}^n} [W(p+q,q)F(x,p+q,t) - W(p,q)F(x,q,t)]\,\mathrm{d}q.$$

7. Quantum equation:

$$\begin{aligned}\frac{\partial F}{\partial t} + \{F,H\} + \sigma F &= \frac{\mathrm{i}}{(2\pi)^n h}\int_{\mathbf{R}^{2n}} \left[\Phi\left(x - \frac{1}{2}hy, t\right)\right.\\ &\left. - \Phi\left(x + \frac{1}{2}hy, t\right)\right] \mathrm{e}^{\mathrm{i}y(p-p')} F(x,p',t)\mathrm{d}p'\mathrm{d}y + \mathrm{St}\,F + q.\end{aligned}$$

Kinetic equations with regard to quantum effects have the form

$$\frac{\partial F}{\partial t} + \{F,H\} + aF = \frac{\mathrm{i}}{(2\pi)^n h}\int_{\mathbf{R}^{2n}} \left[\Phi\left(x - \frac{1}{2}hy, t\right)\right.$$

$$\left. - \Phi\left(x + \frac{1}{2}hy, t\right)\right] \mathrm{e}^{\mathrm{i}y(p-p')} F(x,p',t)\mathrm{d}p'\mathrm{d}y + \mathrm{St}\,F + q(x,p,t). \tag{2.1}$$

Here $F(x,p,t)$ is the quantum distribution function, H is the Hamilton function, Φ is the averaged potential, $a(x,p,t)$ represents the absorption, h is the Planck constant, $\mathrm{St}\,F$ is the collision integral, $q(x,p,t)$ is a function that characterizes the sources.

Given the appropriate information at the boundary, the main direct problems for equations of type (2.1) describe the evolution of an initial distribution. As a rule, all the coefficients in equation (2.1) are considered to be known. Analysis of physical problems connected with kinetic equations (for example problem of tomography) shows that in addition to the distribution function $F(x,p,t)$ certain coefficients contained in these equations should be ascribed to the quantities sought.

The other words some coefficients are unknown in these equations. The problem of determining some functions involved in the kinetic equations, e.g. a or Φ, apart from the distribution function $F(x,p,t)$ is conventionally called the inverse problem for kinetic equations (see Anikonov, 1975; Anikonov, 1978; 1984; 1985; 1986; 1987; 1989; Anikonov and Amirov, 1983; Anikonov and Bubnov, 1988; 1989; Lavrent'ev and Anikonov, 1967; Lavrent'ev et al., 1971). In these problems the main information is the trace of the distribution function on a certain (x,p,t)-manifold, e.g. $x_n = 0$ or $|x| = 1$, etc. Physically, inverse problems for kinetic equations are connected with determining interaction forces, absorptivities, sources, scattering indicatrices and other physical quantities. Some inverse problems of this kind may be formulated as control problems (Anikonov and Bubnov, 1989).

We repeat that our basic information for inverse problems is the function $F_0(s,p,t) = F(x,p,t)|_\Gamma$, $s \in \Gamma$, $p \in D \subset \mathbf{R}^n$, $t_1 \le t \le t_2$, where Γ is a manifold in $\mathbf{R}^n$ of variable x, for example

$$\Gamma = \{x : |x| = 1\} \quad \text{or} \quad \{x : x_n = 0\}.$$

The other words we must know the trace of the distribution function $F(x,p,t)$ on Γ.

The studies concerning predominantly multidimensional inverse problems for kinetic equations suggested and developed by the author (Anikonov, 1978; 1984; 1985; 1986; 1987; 1989; 1991; Anikonov and Amirov, 1983; Anikonov and Bubnov, 1988; 1989; Anikonov and Pestov, 1990) are reported in this work without attempting to survey the whole field.

2.2 AN EXAMPLE OF AN INVERSE PROBLEM FOR KINETIC EQUATION

We consider the inverse problems of determining two functions $(F, q(x,t))$ or $(F, \lambda(x,t))$, $\partial q/\partial p = 0$, $\partial\lambda/\partial p = 0$ in the region $|p| < \infty$, $|x| < a$, $|t - t_0| < b$, $a > 0$, $b > 0$,

$$\frac{\partial F}{\partial t} + p\frac{\partial F}{\partial x} + \lambda(x,t)\frac{\partial F}{\partial p} = q(x,t), \tag{2.2}$$

$$F|_{x=0} = F_0(p,t), \quad |p| < \infty, \quad |t - t_0| < b. \tag{2.3}$$

Equation (2.2) is the one-dimensional kinetic equation (Vlasov equation for $q = 0$) $\lambda(x,t) = -\partial\Phi/\partial x$, and Φ is the potential.

Theorem 2.1. *Suppose that $F_0(p,t)$ is an analytic function in the region $|t - t_0| < \delta$, $|p| < \infty$ while for any t, F_0 is the entire function of exponential type in p which belongs to the space L_2. Then there exist such numbers $a_1 > 0$, $b_1 > 0$ that inverse problems (2.2) and (2.3) to determine (F,q) have an analytical solution in the region $|x| < a_1$, $|t - t_0| < b_1$, $|p| < \infty$.*

Theorem 2.2. *Let the conditions of Theorem 2.1 be met and*

$$\left.\frac{\partial F_0}{\partial p}\right|_{p=0} \neq 0.$$

Then there exist such numbers $a_1 > 0$, $b_1 > 0$ that the inverse problems (2.2), (2.3) to determine (F,λ) have an analytic solution in the region $|x| < a_1$, $|t - t_0| < b_1$, $|p| < \infty$.

2.3 ONE-DIMENSIONAL INVERSE PROBLEMS

In one-dimensional inverse problems explicit inversion formulae can often be obtained for the solution. The following example illustrates this situation. Let us consider the one-dimensional kinetic equation (2.1) with

$$\Phi = 0, \quad \mathrm{St}\, F = 0, \quad q = 0, \quad H = \frac{1}{2}p^2 + \tilde{\Phi}(x).$$

In the case it runs as

$$\frac{\partial F}{\partial x}p - \tilde{\Phi}'(x)\frac{\partial F}{\partial p} + a(x)F = 0. \tag{2.4}$$

Given the information $F_0(p) = F|_{x=0}$, $p \in \mathbf{R}^1$, in equation (2.4), we shall seek the functions $F(x,p)$ and $a(x)$ $(\partial a/\partial p = 0)$. Let us assume that the function $F_0(p) > 0$ is differentiable, and introduce the functions

$$f(p) = \tilde{f}\left(\sqrt{2p}\right), \qquad g(p) = \tilde{g}\left(\sqrt{2p}\right),$$

$$b(p) = \frac{1}{2\pi}\frac{\mathrm{d}}{\mathrm{d}p}\int_0^p \frac{g(z)\mathrm{d}z}{\sqrt{p-z}}, \qquad p \geq 0,$$

where

$$\tilde{f}(p) = \frac{\ln F_0(p) + \ln F_0(-p)}{2}, \qquad \tilde{g}(p) = \frac{\ln F_0(p) - \ln F_0(-p)}{2}.$$

Theorem 2.3. *If the function $F_0(p)$ is continuously differentiable, and $F_0(p) > 0$, $\tilde{\Phi}(x) > 0$, $x > 0$, $\tilde{\Phi}(0) = 0$, then the following formulas hold:*

$$a(x) = 2b\left(\tilde{\Phi}(x)\right)\tilde{\Phi}'(x),$$

$$F(x,p) = \exp\left\{f\left(\frac{1}{2}p^2 + \tilde{\Phi}(x)\right) + \int_{-p}^{p} b\left(\frac{1}{2}p^2 - \frac{1}{2}\eta^2 + \tilde{\Phi}(x)\right)\mathrm{d}\eta\right\}, \quad x \geq 0, \quad p \in \mathbf{R}^1.$$

REMARK 2.1. In addition, assume that the functions $\tilde{\Phi}(x)$ and $F(x,p)$ satisfy the relation

$$\tilde{\Phi}''(x) = 4\pi\sigma\int_{-\infty}^{\infty} F(x,p)\mathrm{d}p.$$

Then according to the formulas of Theorem 2.3 under appropriate restrictions on $F_0(p)$ we obtain the equation for the potential $\tilde{\Phi}(x)$

$$\tilde{\Phi}''(x) = P(\tilde{\Phi}),$$

where function P is known and depend only on the initial information $F_0(p)$ by the formula

$$P(z) = 4\pi\sigma\int_{-\infty}^{\infty} \exp\left\{f\left(\frac{1}{2}p^2 + z\right) + \int_{-p}^{p} b\left(\frac{1}{2}p^2 - \frac{1}{2}\eta^2 + z\right)\mathrm{d}\eta\right\}\mathrm{d}p.$$

In view of this, given $\tilde{\Phi}(0)$, $\tilde{\Phi}(0)$ and $\tilde{\Phi}'(0)$ the function $\tilde{\Phi}(x)$ may be considered as the sought function along with a and F. Here we omit the formulation of results.

2.4 MULTIDIMENSIONAL INVERSE PROBLEMS

This section deal with the multidimensional version of the inverse problem considered in section 2.1. In the multidimensional case the unknown coefficient also turns out to be found by an explicit formula used as a solution of the Cauchy problem for an equation of the evolution type with respect to a particular spatial variable. Here we also restrict ourselves to one example. Let $H = \frac{1}{2}p^2$ in equation (2.1), a and F be the sought functions, and the operator St be given. We define the operator A by the formula

$$AF = \frac{\partial F}{\partial t} + \sum_{j=1}^{n-1} \frac{\partial F}{\partial x_j} p_j - \frac{\mathrm{i}}{(2\pi)^n h} \int\limits_{\mathbf{R}^{2n}} [\Phi]\, \mathrm{e}^{\mathrm{i}y(p-p')} F(x,p',t) \mathrm{d}p' \mathrm{d}y - \mathrm{St}\, F - q.$$

Let the function

$$F_0(x',p,t) = F|_{x_n}, \qquad x' = (x_1, x_2, \ldots, x_{n-1})$$

be given, and

$$\frac{\partial a}{\partial p_n} = 0.$$

Note that the operator A does not involve differentiation with respect to x_n, it is independent of a and known, if Φ and St are given.

Theorem 2.4. *Let a twice differentiable function $F > 0$ be the solution to the Cauchy problem for the evolution equation with respect to the variable x_n:*

$$\frac{\partial F}{\partial x_n} = F \int\limits_0^1 \frac{\partial}{\partial p_n} \left[\frac{1}{F} AF\right] (zp_n) \mathrm{d}z, \qquad F|_{x_n=0} = F_0(x',p,t).$$

Then the formula

$$a = \left(AF + \frac{\partial F}{\partial x_n}\right) / F$$

is valid.

REMARK 2.2. Under the condition

$$\Delta\tilde{\Phi} = 4\pi\sigma \int\limits_{\mathbf{R}^n} F(x,p,t) \mathrm{d}p.$$

and the boundary condition for $\Phi(x,t)$, the latter may be considered as the sought function, along with a and F, similar to the case above. In classes of entire functions the solvability of the Cauchy problem similar to that of Theorem 2.4 may be studied on the basis of Banach space scales (see Lavrent'ev et al., 1971).

Some results in the analytic case.

In the Euclidean space $\mathbf{R}^{2n+1}$, $n \geq 2$ we consider the kinetic equations

$$\frac{\partial F}{\partial t} + \sum_{j=1}^{n} \left(\frac{\partial F}{\partial x_j} p_j + \frac{\partial F}{\partial p_j} \frac{\partial \Phi}{\partial x_j} \right) = \mathrm{St}\, F. \tag{2.5}$$

Below it will be convenient for us to represent the collision integral St F as follows: St $F = aF + b$ where a and b are functions of variables (x,p,t) which are possibly generated by the distribution $F(x,p,t)$.

Let

$$\varphi = \arctan\left(\frac{p_n}{p_{n-1}}\right), \quad q = \sqrt{p_n^2 + p_{n-1}^2}, \quad x = x_{n-1}, \quad y = x_n,$$

$$LF = \sum_{j=1}^{n-2}\left(\frac{\partial F}{\partial x_j}p_j + \frac{\partial F}{\partial p_j}\frac{\partial \Phi}{\partial x_j}\right).$$

With these variables (2.5) can be written as follows:

$$\frac{\partial F}{\partial t} + q\left(\frac{\partial F}{\partial x}\cos\varphi + \frac{\partial F}{\partial y}\sin\varphi\right) + \frac{\partial F}{\partial q}\left(\frac{\partial \Phi}{\partial x}\cos\varphi + \frac{\partial \Phi}{\partial y}\sin\varphi\right)$$

$$-\frac{1}{q}\frac{\partial F}{\partial \varphi}\left(\frac{\partial \Phi}{\partial y}\cos\varphi - \frac{\partial \Phi}{\partial x}\sin\varphi\right) + LF = aF + b. \tag{2.6}$$

Below it is assumed that the functions Φ, a and b in (2.6) do not depend on the variable φ, i.e.

$$\frac{\partial \Phi}{\partial \varphi} = 0, \quad \frac{\partial a}{\partial \varphi} = 0, \quad \frac{\partial b}{\partial \varphi} = 0.$$

We formulate the following inverse problems:

Problem I. Find the functions F, a and b if the functions $F_0 = F|_{y=0}$ and Φ are known.

Problem II. Find the functions F, Φ and b if the functions $\Phi_0 = \Phi|_{y=0}$ and $F_0 = F|_{y=0}$ are known.

Problem III. Find the functions F, Φ and a if the functions $F_0 = F|_{y=0}$, $\Phi_0 = \Phi|_{y=0}$ and b are known.

All functions considered below are assumed to be analytic in the neighbourhood of the point $(0, q_0)$, $q_0 > 0$ of the space $\mathbf{R}^{2n+1}$, and periodic in φ with period 2π.

Theorem 2.5. *Suppose that for each inverse problems I–III the conditions of the same number below is satisfied.*

1. $F_0|_{\varphi=0} - F_0|_{\varphi=\pi} \neq 0$,

2. $\left.\frac{\partial F_0}{\partial \varphi}\right|_{\varphi=0} + \left.\frac{\partial F_0}{\partial \varphi}\right|_{\varphi=\pi} \neq 0$,

3. $\left.\frac{\partial F_0}{\partial \varphi}F_0\right|_{\varphi=0} + \left.\frac{\partial F_0}{\partial \varphi}F_0\right|_{\varphi=\pi} \neq 0$.

Then the uniqueness of the solution of these problems is guaranteed in the class of analytic functions.

Theorem 2.6. *Suppose that*

$$F_0 = \sum_{k=0}^{\infty} u_k^0 \cos k\varphi + v_k^0 \sin k\varphi,$$

where $|u_k^0| < Ae^{-pk}$, $|v_k^0| < Ae^{-pk}$, $k = 0, 1, 2\ldots$, $A > 0$, $p > 0$ *are constants and conditions 1–3 of Theorem 2.5 are satisfied. Then there exist analytic solutions of inverse problems I–III.*

Consider the inverse problem for kinetic equation in the multidimensional case:

$$\frac{\partial F}{\partial t} + q\frac{\partial F}{\partial y} - \frac{\partial F}{\partial q}a + \{F, H\} = \sigma F + \mathrm{St}\, F + f;$$

find $F(x, y, q, p, t)$, $\sigma(x, p, t)$, if function $F|_{y=0} = F_0(x, q, p, t)$ is known (a, f and St *are known also,* St F *is Boltzman the collision integral).*

Lemma 2.1. *Function $F(x, y, q, p, t)$ is the solution of problem*

$$\frac{\partial F}{\partial y} = F\int_0^1 \frac{\partial}{\partial q}\left\{\frac{1}{F}\left[F + \frac{\partial F}{\partial q}a - \{F, H\} + \mathrm{St}\, F + F\right]\right\}(q\cdot\eta)\mathrm{d}\eta, \quad F|_{y=0} = F_0.$$

Theorem 2.7. *In the analytic case of the inverse problem of finding $F > 0$, σ has no more than one solution.*

Let us consider the uniqueness of determining the potential $\Phi(x, t)$ from equation (2.1) using the functions $a_k(p, t)$, $b_k(p, t)$, $k = 1, 2$, $p \in \mathbf{R}^n$, $\alpha \le t \le \beta$, which are defined by the relations

$$a_k(p, t) = F|_{x=x_k^0}, \quad b_k(p, t) = \sum_{j=1}^n \frac{\partial F}{\partial x_j}p_j\bigg|_{x=x_k^0} \tag{2.7}$$

where x_1^0, x_2^0 are fixed points of the domain $D \subset \mathbf{R}^n$. Assume that $c_k(p, t) = \mathrm{St}\, a_k$, and that there exist Fourier transforms $\hat{\Phi}$, $\hat{a}_k$, $\hat{b}_k$, $\hat{c}_k$ of the functions Φ, a_k, b_k, c_k with respect to (x, p) respectively. It turns out that the presence of the quantum term in (2.1) allows us to formulate the following result.

Theorem 2.8. *Assume that the functions*

$$A_k(z, t) = \frac{h^2}{4}\int_{\mathbf{R}^n} \frac{(\partial\hat{a}_k/\partial t) + \hat{b}_k - \hat{c}_k}{\hat{a}_k}\mathrm{e}^{-\mathrm{i}zy}\mathrm{d}y, \quad k = 1, 2$$

are defined correctly. Then the formula

$$\hat{\Phi}\left(\frac{2z}{h}, t\right) = \frac{A_1\,\mathrm{e}^{-\mathrm{i}(x_2^0, z)} - A_2\,\mathrm{e}^{-\mathrm{i}(x_1^0, z)}}{\sin 2(x_1^2 - x_2^2, z)}$$

holds.

REMARK 2.3. *The additional information (2.7) permits posing the problem of simultaneously determining three functions F, a, Φ from (2.1) using Theorem 2.4 and 2.8.*

2.5 A UNIQUENESS THEOREM FOR THE SOLUTION OF AN INVERSE PROBLEM FOR A KINETIC EQUATION

Let Q be a domain in the real Euclidean space $\mathbf{R}^{2n+1}$, $n \ge 1$, of the variables $(\bar{x}, \bar{p}, t)$ where $\bar{x} \in D$ and D is a domain in $\mathbf{R}^n$ with a smooth boundary while the variables $\bar{p} \in \mathbf{R}^n$ and $t \in \mathbf{R}$ are subject to the inequalities $|p_i - p_i^0| < a_i$ and $|t - t_0| < b$, where $a_i > 0$, p_i^0, $i = 1, \ldots, n$, $b > 0$, and t_0 are fixed numbers.

In Q we consider the kinetic equation

$$\frac{\partial W}{\partial t}+\sum_{i=1}^{n}\left(\frac{\partial W}{\partial x_i}\frac{\partial H}{\partial p_i}-\frac{\partial W}{\partial p_i}\frac{\partial H}{\partial x_i}\right)=\lambda(\bar{x},\bar{p},t)W \tag{2.8}$$

with on the right side $\lambda(\bar{x},\bar{p},t)$, satisfying the equation

$$\sum_{j=1}^{n}\frac{\partial^2\lambda}{\partial p_j\partial x_j}=0.$$

We pose the inverse problem: find functions $W(\bar{x},\bar{p},t)$ and $\lambda(\bar{x},\bar{p},t)$ in Q if the Hamiltonian $H(\bar{x},\bar{p})\in C^2(\bar{Q})$ is given and the trace of the solution $W(\bar{x},\bar{p},t)$ of (2.8) on the boundary Γ of Q is known, i.e. $W|_\Gamma=W_0(\bar{x},\bar{p},t)>0$, $(\bar{x},\bar{p},t)\in\Gamma$, where W_0 is a known function, $W_0>0$.

Theorem 2.9. *If in the domain Q the matrices $(\partial^2 H/\partial p_i\partial p_j)$ and $(-\partial^2 H/\partial x_i\partial x_j)$, $i,j=1,\dots,n$, are positive definite, then the inverse problem has no more than one solution $F(\bar{x},\bar{p},t)\in C^2(\bar{Q})$, $\lambda(\bar{x},\bar{p},t)\in C^2(\bar{Q})$.*

We sketch the proof. Let $F=\ln W$, then we have a linear inverse problem:

$$\frac{\partial F}{\partial t}+\sum_{i=1}^{n}\left(\frac{\partial F}{\partial x_i}\frac{\partial H}{\partial p_i}-\frac{\partial F}{\partial p_i}\frac{\partial H}{\partial x_i}\right)=\lambda,$$

$$F\,|_\Gamma=F_0=\ln W_0.$$

Suppose that $F(\bar{x},\bar{p},t)$, $\lambda(\bar{x},\bar{p},t)$ is a solution of the inverse problem such that $F\,|_\Gamma=0$. We shall show that $F(\bar{x},\bar{p},t)=0$ and $\lambda(\bar{x},\bar{p},t)=0$ in Q. By the hypothesis of the problem the function $\lambda(\bar{x},\bar{p},t)$ satisfies the equation

$$\sum_{1}^{n}\frac{\partial^2\lambda}{\partial p_j\partial x_j}=0.$$

Hence

$$\sum_{j=1}^{n}\frac{\partial F}{\partial x_j}\left[\frac{\partial}{\partial p_j}\left(\frac{\partial F}{\partial t}+\sum_{i=1}^{n}\left(\frac{\partial F}{\partial x_i}\frac{\partial H}{\partial p_i}-\frac{\partial F}{\partial p_i}\frac{\partial H}{\partial x_i}\right)\right)\right]=\sum_{j=1}^{n}\frac{\partial F}{\partial x_j}\frac{\partial\lambda}{\partial p_j}=\sum_{j=1}^{n}\frac{\partial}{\partial x_j}\left(F\frac{\partial\lambda}{\partial p_j}\right).$$

It turns out that

$$\begin{aligned}
&\frac{1}{2}\sum_{i,j=1}^{n}\left(\frac{\partial^2 H}{\partial p_j\partial p_j}\frac{\partial F}{\partial x_i}\frac{\partial F}{\partial x_j}-\frac{\partial^2 H}{\partial x_j\partial x_j}\frac{\partial F}{\partial p_i}\frac{\partial F}{\partial p_j}\right)+\frac{1}{2}\sum_{i,j=1}^{n}\frac{\partial}{\partial p_j}\left[\frac{\partial F}{\partial x_j}\left(\frac{\partial F}{\partial x_i}\frac{\partial H}{\partial p_i}-\frac{\partial F}{\partial p_i}\frac{\partial H}{\partial x_i}\right)\right]\\
&-\frac{1}{2}\sum_{i,j=1}^{n}\frac{\partial}{\partial x_j}\left[\frac{\partial F}{\partial p_j}\left(\frac{\partial F}{\partial x_i}\frac{\partial H}{\partial p_i}-\frac{\partial F}{\partial p_i}\frac{\partial H}{\partial x_i}\right)\right]+\frac{1}{2}\sum_{i,j=1}^{n}\frac{\partial}{\partial x_i}\left(\frac{\partial H}{\partial p_i}\frac{\partial F}{\partial x_j}\frac{\partial F}{\partial p_j}\right)\\
&-\frac{1}{2}\sum_{i,j=1}^{n}\frac{\partial}{\partial p_i}\left(\frac{\partial H}{\partial x_i}\frac{\partial F}{\partial x_j}\frac{\partial F}{\partial p_j}\right)+\frac{1}{2}\sum_{j=1}^{n}\left[\frac{\partial}{\partial t}\left(\frac{\partial F}{\partial x_j}\frac{\partial F}{\partial p_j}\right)\right.\\
&\left.-\frac{\partial}{\partial x_j}\left(\frac{\partial F}{\partial t}\frac{\partial F}{\partial p_j}\right)+\frac{\partial}{\partial p_j}\left(\frac{\partial F}{\partial t}\frac{\partial F}{\partial x_j}\right)\right]=\sum_{j=1}^{n}\frac{\partial}{\partial x_j}\left(F\frac{\partial\lambda}{\partial p_j}\right).
\end{aligned} \tag{2.9}$$

Integrating both sides of (2.9) over Q and considering the condition $F|_\Gamma = 0$, we obtain

$$\int_Q \sum_{i,j=1}^{n} \left(\frac{\partial^2 H}{\partial p_j \partial p_j} \frac{\partial F}{\partial x_i} \frac{\partial F}{\partial x_j} - \frac{\partial^2 H}{\partial x_j \partial x_j} \frac{\partial F}{\partial p_i} \frac{\partial F}{\partial p_j} \right) d\bar{x} d\bar{p} dt = 0. \tag{2.10}$$

By the hypothesis of the theorem the matrices $(\partial^2 H/\partial p_i \partial p_j)$ and $(-\partial^2 H/\partial x_i \partial x_j)$ are positive definite. Therefore it follows from (2.10) that $\partial F/\partial x_i = 0$ and $\partial F/\partial p_i = 0$, $i = 1, \ldots, n$ which with $F|_\Gamma = 0$ give the equality $F(\bar{x}, \bar{p}, t) = 0$, $(\bar{x}, \bar{p}, t) \in Q$. From (2.8) it then follows that $\lambda(\bar{x}, \bar{p}, t) = 0$.

2.6 THE GENERAL UNIQUENESS THEOREM

We consider the simultaneous determination of three functions (F, H, λ) from (2.1) where F is the distribution function, H is the Hamilton function, and $\lambda = \operatorname{St} F + q$. The uniqueness of the solution to the inverse problem can be proved under special restriction concerning both the data and the sought functions (F, H, λ). Let $D \subset \mathbf{R}^n$ be a domain of variables p with a smooth boundary; $Q = D \times D_1$, $x \in D$, and ∂Q be the boundary of Q. The data for the inverse problems are the functions F_0, A, B, C, H_0 defined as follows:

$$\begin{gathered} F_0(s, p, t) = F|_\Gamma, \quad s \in \Gamma = \partial D, \quad p \in \mathbf{R}^n, \quad \alpha \le t \le \beta, \\ A(x, p) = F|_{t=\alpha}, \quad B(x, p) = F|_{t=\beta}, \quad C(x, p) = \left.\tfrac{\partial F}{\partial t}\right|_{t=\alpha}, \quad H_0 = H|_{\partial Q}. \end{gathered} \tag{2.11}$$

Suppose that the following conditions are satisfied.

1. The sought Hamilton function $H(x, p)$ is analytic in $p \in \mathbf{R}^n$ and $H(x, p) \in C^2(\bar{D} \times \mathbf{R}^n)$.
2. $\sum_{i,j=1}^{n} \frac{\partial^2 H}{\partial x_i \partial x_j} \xi_i \xi_j \le 0, \quad \sum_{i,j=1}^{n} \frac{\partial^2 H}{\partial p_i \partial p_j} \xi_i \xi_j \ge \alpha |\xi|^2, \quad \forall \xi \in \mathbf{R}^n$, with constant $\alpha > 0$.
3. $|\mathbf{D}^\alpha H(x, p)| \le (N + |p|)^m$, $|\alpha| \le 2$, $N > 0$, $m > 0$ are constants, $\mathbf{D}^\alpha$ being a differentiation operator with respect to variables (x, p).
4. The known potential $\Phi(x, t)$ belongs to the class $C^2(\mathbf{R}^n \times [a, b])$, $p \in \mathbf{R}^n$, $\alpha \le t \le \beta$.
5. $|D_x^\gamma \Phi| \le (N_1 + |p|)^{m_1}$, $|\gamma| \le 2$, $N_1 > 0$, $m_1 > 0$ are constants.
6. $\sum_{i,j=1}^{n} \frac{\partial^2 \Phi}{\partial x_i \partial x_j} \xi_i \xi_j \le 0, \quad \forall \xi \in \mathbf{R}^n$.
7. The given function obeys the conditions
$$A \in C^2(\bar{Q}), \quad \sum_{i,j=1}^{n} \frac{\partial^2 A}{\partial x_i \partial x_j} \xi_i \xi_j \ge 0, \quad \sum_{i,j=1}^{n} \frac{\partial^2 A}{\partial p_i \partial p_j} \xi_i \xi_j < 0, \quad \forall \bar{\xi} \ne 0.$$
8. The sought quantum distribution function $F(x, p, t) \in C^2(\bar{D} \times \mathbf{R}^n \times [a, b])$ together with its first and second derivatives, rapidly vanishes by
$$\lim_{|p| \to \infty} |D^\alpha F|\, |p|^m = 0, \quad \forall m > 0, \quad |\alpha| \le 2.$$

9. The unknown function $q(x,p,t)$ is twice and continuously differentiable and satisfies the equation

$$\sum_{j=1}^{n} \frac{\partial^2 q}{\partial x_j \partial p_j} = 0.$$

10. The operator

$$\mathbf{M} = \sum_{j=1}^{n} \frac{\partial^2}{\partial x_j \partial p_j} \operatorname{St}$$

is supposed to be known, defined on the set of distribution functions and to map this set onto itself.

11. For any two distribution functions $F_1(x,p,t)$ and $F_2(x,p,t)$, $F_1 \neq F_2$, the following inequality holds

$$\int_a^b \int_D \int_{\mathbf{R}^n} \left\{ \frac{\alpha}{2} \left|\operatorname{grad}_x (F_1 - F_2)\right|^2 + (F_1 - F_2)(MF_1 - MF_2) \right\} dx dp dt > 0.$$

Let us make some remarks concerning the conditions stated. Conditions 2, 6 and 7 imply the convexity of the functions $H(x,p)$, $A(x,p)$ and $\Phi(x,p)$ with respect to the corresponding variables. In the case $n = 1$ these functions may be as follows

$$H = \frac{1}{2}p^2, \quad p \in \mathbf{R}^1, \quad A = \mathrm{e}^{-p^2}, \quad |p| < \frac{1}{\sqrt{2}},$$

$$\Phi(x,t) = -\gamma(t)x^2 + \beta(t)x + c(t), \quad \gamma > 0.$$

Conditions 3, 5 and 8 characterize an increase and a decrease, respectively, in functions $H(x,p)$, $\Phi(x,p)$ and $F(x,p,t)$ at infinity. Condition 8 is usually considered to be satisfied. Sometimes the sources $q(x,p,t)$ are independent of p or x. These cases are embraced by condition 9. Items 10 and 11 state the restrictions concerning the aprioriunknown operator St. Condition 11 is connected with the monotonicity of the operator

$$\mathbf{M} = \sum_{j=1}^{n} \frac{\partial^2}{\partial x_j \partial p_j} \operatorname{St}$$

which, unlike operator St is considered as given.

The inverse problem is: find the functions (F, H, λ), given (2.11) for equation (2.1) and provided that conditions 1–11 are satisfied.

Theorem 2.10. *The inverse problem has no more than one solution.*

PROOF. We first prove that the Hamiltonian $H(x,p)$, $x \in D$, $p \in \mathbf{R}^n$ is uniquely determined.

Setting $t = 0$ in (2.1), we have

$$C(x,p) + \{A, H\} = \frac{\mathrm{i}}{(2\pi)^n h} \int_{\mathbf{R}^{2n}} [\Phi]_0 \, \mathrm{e}^{\,iy(p-p')} A(x,p') dp' dy + \operatorname{St} F + q_0(x,p). \tag{2.12}$$

Here

$$q_0 = q(x,p,0), \quad [\Phi]_0 = \left[\Phi\left(x - \frac{1}{2}hy, 0\right) - \Phi\left(x + \frac{1}{2}hy, 0\right)\right].$$

If H_1 and H_2 are two solutions of (2.12) satisfying $H_j|_{\partial Q} = H_0$, we consider the difference $\varphi = H_1 - H_2$. We have

$$\{\varphi, A\} = \mathrm{St}_2\, A - \mathrm{St}_2\, A + q_{01} - q_{02} = \lambda_0, \quad \varphi|_{\partial Q} = 0.$$

The right-hand side $\mathrm{St}_2\, A - \mathrm{St}_2\, A + q_{01} - q_{02} = \lambda_0(x,p)$ can be shown to satisfy

$$\sum_{j=0}^{n} \frac{\partial^2 \lambda_0}{\partial x_j \partial p_j} = 0.$$

Indeed, by conditions 9 and 10

$$\sum_{j=1}^{n} \frac{\partial^2}{\partial x_j \partial p_j} [(\mathrm{St}_2\, A - \mathrm{St}_2\, A) + (q_{01} - q_{02})] = \mathbf{M}A - \mathbf{M}A = 0.$$

Thus,

$$\{\varphi, A\} = \lambda_0, \quad \sum_{j=0}^{n} \frac{\partial^2 \lambda_0}{\partial x_j \partial p_j} = 0, \quad \varphi\,|_{\partial Q} = 0. \tag{2.13}$$

Using (2.13) one can verify directly that

$$\begin{aligned}\sum_{j=1}^{n} \frac{\partial \varphi}{\partial x_j} \frac{\partial}{\partial p_j} \{\varphi, A\} &= \frac{1}{2} \sum_{i,j=1}^{n} \left(\frac{\partial^2 A}{\partial p_i \partial p_j} \frac{\partial \varphi}{\partial x_i} \frac{\partial \varphi}{\partial x_j} - \frac{\partial^2 A}{\partial x_i \partial x_j} \frac{\partial \varphi}{\partial p_i} \frac{\partial \varphi}{\partial p_j} \right) \\ &+ \frac{1}{2} \sum_{j=1}^{n} \frac{\partial \varphi}{\partial p_j} \frac{\partial}{\partial x_j} \frac{\partial}{\partial p_j} \{\varphi, A\} - \frac{1}{2} \sum_{j=1}^{n} \frac{\partial}{\partial x_j} \frac{\partial}{\partial p_j} \{\varphi, A\} \\ &+ \frac{1}{2} \sum_{i,j=1}^{n} \frac{\partial}{\partial x_i} \left(\frac{\partial A}{\partial p_i} \frac{\partial \varphi}{\partial x_j} \frac{\partial \varphi}{\partial p_j} \right) - \frac{1}{2} \sum_{i,j=1}^{n} \frac{\partial}{\partial p_i} \left(\frac{\partial A}{\partial x_i} \frac{\partial \varphi}{\partial x_j} \frac{\partial \varphi}{\partial p_j} \right) \\ &= \sum_{j=1}^{n} \frac{\partial}{\partial x_j} \left(\varphi \frac{\partial \lambda_0}{\partial p_j} \right).\end{aligned}$$

Recalling the condition $\varphi|_{\partial Q} = 0$ and making use of the essential fact that the domain Q is a Cartesian product $D \times D_1$, $x \in D$, $p \in D_1$, we then obtain by integration

$$\int_Q \sum_{i,j=1}^{n} \left(\frac{\partial^2 A}{\partial p_i \partial p_j} \frac{\partial \varphi}{\partial x_i} \frac{\partial \varphi}{\partial x_j} - \frac{\partial^2 A}{\partial x_i \partial x_j} \frac{\partial \varphi}{\partial p_i} \frac{\partial \varphi}{\partial p_j} \right) \mathrm{d}x \mathrm{d}p = 0.$$

From this and condition 7 we obtain $\partial\varphi/\partial x_j$ for $j = 1, \ldots, n$, which in turn leads to the relation $\varphi(x,p) = 0$, $(x,p) \in Q$. Thus, $H_1 = H_2$ in Q. Since by condition 1 the Hamiltonian is analytic, it has a unique continuation from the domain Q to $D \times \mathbf{R}^n$, on which it will henceforth be taken as known.

Let (F_1, λ_1) and (F_2, λ_2) be two solutions of the inverse problem satisfying all the hypotheses of the theorem. Writing $F = F_1 - F_2$ and $q = q_1 - q_2$, from (2.1) and (2.12) we obtain

$$\frac{\partial F}{\partial t} + \{F, H\} = \frac{\mathrm{i}}{(2\pi)^n h} \int_{\mathbf{R}^{2n}} [\Phi]\, \mathrm{e}^{\mathrm{i}y(p-p')} F(x,p',t) \mathrm{d}p' \mathrm{d}y + \mathrm{St}_1\, F_1 + \mathrm{St}_2\, F_2 + q, \tag{2.14}$$

$$F|_{t=a} = 0, \quad F|_{t=b} = 0, \quad F|_{x\in\Gamma} = 0.$$

Differentiating (2.14) with respect to p_j, multiplying by $\partial F/\partial x_j$, and summing over j, we get the equality

$$\sum_{j=1}^{n}\frac{\partial F}{\partial x_j}\frac{\partial}{\partial p_j}\left(\frac{\partial F}{\partial t}+\{F,H\}\right)$$

$$=\sum_{j=1}^{n}\frac{\partial F}{\partial x_j}\left\{\frac{\partial}{\partial p_j}\frac{\mathrm{i}}{(2\pi)^n h}\right\}\int_{\mathbf{R}^{2n}}[\Phi]\,\mathrm{e}^{\mathrm{i}y(p-p')}F(x,p',H)\mathrm{d}p'\mathrm{d}y+\mathrm{St}_1\,F_1+\mathrm{St}_2\,F_2+q. \tag{2.15}$$

The following relations can be verified directly:

$$\frac{1}{2}\sum_{j=1}^{n}\frac{\partial}{\partial t}\left(\frac{\partial F}{\partial x_j}\frac{\partial F}{\partial p_j}\right)-\frac{1}{2}\frac{\partial}{\partial x_j}\left(\frac{\partial F}{\partial t}\frac{\partial F}{\partial p_j}\right)+\frac{1}{2}\frac{\partial}{\partial p_j}\left(\frac{\partial F}{\partial t}\frac{\partial F}{\partial x_j}\right)=\sum_{j=1}^{n}\frac{\partial F}{\partial x_j}\frac{\partial^2 F}{\partial p_j\partial t}, \tag{2.16}$$

$$\sum_{j=1}^{n}\frac{\partial F}{\partial x_j}\frac{\partial}{\partial p_j}\{F,H\}=\frac{1}{2}\sum_{i,j=1}^{n}\left[\frac{\partial^2 H}{\partial p_i\partial p_j}\frac{\partial F}{\partial x_i}\frac{\partial F}{\partial x_j}-\frac{\partial^2 H}{\partial x_i\partial x_j}\frac{\partial F}{\partial p_i}\frac{\partial F}{\partial p_j}\right]$$

$$+\frac{1}{2}\sum_{i,j=1}^{n}\frac{\partial}{\partial p_j}\left[\frac{\partial F}{\partial x_j}\left(\frac{\partial F}{\partial x_i}\frac{\partial H}{\partial p_i}-\frac{\partial F}{\partial p_i}\frac{\partial H}{\partial x_i}\right)\right]-\frac{1}{2}\sum_{i,j=1}^{n}\frac{\partial}{\partial x_j}\left[\frac{\partial F}{\partial p_j}\left(\frac{\partial F}{\partial x_i}\frac{\partial H}{\partial p_i}-\frac{\partial F}{\partial p_i}\frac{\partial H}{\partial x_i}\right)\right]$$

$$+\frac{1}{2}\sum_{i,j=1}^{n}\frac{\partial}{\partial x_i}\left(\frac{\partial H}{\partial p_i}\frac{\partial F}{\partial x_j}\frac{\partial F}{\partial p_j}\right)-\frac{1}{2}\sum_{i,j=1}^{n}\frac{\partial}{\partial p_i}\left(\frac{\partial H}{\partial x_i}\frac{\partial F}{\partial x_j}\frac{\partial F}{\partial p_j}\right). \tag{2.17}$$

Since by hypothesis

$$\sum_{j=1}^{n}\frac{\partial^2 q}{\partial x_j\partial p_j}=0\quad\text{and}\quad\mathbf{M}=\sum_{j=1}^{n}\frac{\partial^2}{\partial x_j\partial p_j}\mathrm{St}$$

we have the equality

$$\sum_{j=1}^{n}\frac{\partial F}{\partial x_j}\frac{\partial}{\partial p_j}(\mathrm{St}\,F_1-\mathrm{St}\,F_2+q)$$

$$=\sum_{j=1}^{n}\frac{\partial}{\partial x_j}F\frac{\partial}{\partial p_j}(\mathrm{St}\,F_1-\mathrm{St}\,F_2+q)-(F_1-F_2)(MF_1-MF_2). \tag{2.18}$$

We now establish that

$$\int_a^b\int_D\int_{\mathbf{R}^n}\sum_{j=1}^{n}\frac{\partial F}{\partial x_j}\frac{\partial}{\partial p_j}\frac{\mathrm{i}}{(2\pi)^n h}\int_{\mathbf{R}^{2n}}[\Phi]\,\mathrm{e}^{\mathrm{i}y(p-p')}F\mathrm{d}p'\mathrm{d}x\mathrm{d}t$$

$$=\frac{1}{2h}\int_a^b\int_D\int_{\mathbf{R}^n}\sum_{j=1}^{n}\frac{\partial}{\partial x_j}\left|\bar{F}\right|^2 y_j[\Phi]\mathrm{d}y\mathrm{d}x\mathrm{d}t, \tag{2.19}$$

where

$$\bar{F}=\frac{1}{(2\pi)^{n/2}}\int_{\mathbf{R}^n}F(x,p',t)\,\mathrm{e}^{\mathrm{i}yp'}\mathrm{d}p'.$$

We have

$$\sum_{j=1}^{n}\frac{\partial F}{\partial x_j}\frac{\partial}{\partial p_j}\frac{\mathrm{i}}{(2\pi)^n h}\int_{\mathbf{R}^{2n}}\left[\Phi\left(x-\frac{1}{2}hy,t\right)-\Phi\left(x+\frac{1}{2}hy,t\right)\right]\mathrm{e}^{\mathrm{i}y(p'-p)}\times F(x,p',t)\mathrm{d}p'\mathrm{d}y$$

$$= \sum_{j=1}^{n} \frac{\partial F}{\partial x_j} \frac{\mathrm{i}}{(2\pi)^n h} \int\limits_{\mathbf{R}^{2n}} [\Phi] y_j \, \mathrm{e}^{\mathrm{i}yp'} e^{-\mathrm{i}yp} F(x,p',t) \mathrm{d}p' \mathrm{d}y$$

$$= \sum_{j=1}^{n} \frac{\partial F}{\partial x_j} \frac{1}{(2\pi)^{n/2} h} \int\limits_{\mathbf{R}^{n}} [\Phi] y_j \bar{F}(x,y,t) \, \mathrm{e}^{-\mathrm{i}yp} \mathrm{d}y$$

whence by integration we obtain

$$\int\limits_a^b \int\limits_D \int\limits_{\mathbf{R}^n} \sum_{j=1}^{n} \frac{\partial F}{\partial x_j} \frac{1}{h} \frac{1}{(2\pi)^{n/2}} \int\limits_{\mathbf{R}^n} [\Phi] y_j \bar{F} \, \mathrm{e}^{-\mathrm{i}yp} \mathrm{d}y \mathrm{d}p \mathrm{d}x \mathrm{d}t$$

$$= \frac{1}{h} \int\limits_a^b \int\limits_D \int\limits_{\mathbf{R}^n} \sum_{j=1}^{n} \frac{\partial \bar{F}^*(x,y,t)}{\partial x_j} \bar{F}(x,y,t) [\Phi] y_j \mathrm{d}y \mathrm{d}x \mathrm{d}t,$$

where

$$\bar{F}^*(x,y,t) = \frac{1}{(2\pi)^{n/2}} \int\limits_{\mathbf{R}^n} F(x,p,t) \, \mathrm{e}^{-\mathrm{i}yp} \mathrm{d}p.$$

We prove next that

$$\frac{1}{h} \int\limits_a^b \int\limits_D \int\limits_{\mathbf{R}^n} \sum_{j=1}^{n} \frac{\partial \bar{F}^*}{\partial x_j} \bar{F} [\Phi] y_j \mathrm{d}y \mathrm{d}x \mathrm{d}t = \frac{1}{2h} \int\limits_a^b \int\limits_D \int\limits_{\mathbf{R}^n} \sum_{j=1}^{n} \frac{\partial}{\partial x_j} \left|\bar{F}\right|^2 y_j [\Phi] \mathrm{d}y \mathrm{d}x \mathrm{d}t. \tag{2.20}$$

Let

$$\bar{F} = \frac{1}{(2\pi)^{n/2}} \int\limits_{\mathbf{R}^n} F(x,p,t) \, \mathrm{e}^{\mathrm{i}yp} \mathrm{d}p = P(x,y,t) + \mathrm{i}Q(x,y,t).$$

Since the function F is real-valued, P and Q are even and odd in the variable y, respectively. We have

$$\int\limits_a^b \int\limits_D \int\limits_{\mathbf{R}^n} \sum_{j=1}^{n} \frac{\partial \bar{F}^*}{\partial x_j} \bar{F} [\Phi] y_j \mathrm{d}y \mathrm{d}x \mathrm{d}t$$

$$= \int\limits_a^b \int\limits_D \int\limits_{\mathbf{R}^n} \left[\sum_{j=1}^{n} \frac{\partial}{\partial x_j} \left(P^2 + Q^2\right) + \mathrm{i} \left(\frac{\partial P}{\partial x_j} Q - P \frac{\partial Q}{\partial x_j} \right) \right] [\Phi] y_j \mathrm{d}y \mathrm{d}x \mathrm{d}t.$$

Since the function $[\Phi]$ is odd in y, the same is true of the function

$$[\Phi] y_j \left(\frac{\partial P}{\partial x_j} Q - P \frac{\partial Q}{\partial x_j} \right),$$

for all j, $1 \le j \le n$, and we have

$$\int\limits_{\mathbf{R}^n} \left(\frac{\partial P}{\partial x_j} Q - P \frac{\partial Q}{\partial x_j} \right) [\Phi] y_j \mathrm{d}y = 0.$$

This establishes (2.20).

We now show that

$$\sum_{j=1}^{n} y_j \frac{\partial}{\partial x_j} \left[\Phi\left(x - \frac{1}{2} hy, t\right) - \Phi\left(x + \frac{1}{2} hy, t\right) \right] \ge 0. \tag{2.21}$$

Indeed,

$$\sum_{j=1}^{n} y_j \frac{\partial}{\partial x_j}[\Phi] = \sum_{j=1}^{n} y_j \left[\frac{\partial \Phi\left(x - \frac{1}{2}hy, t\right)}{\partial x_j} - \frac{\partial \Phi\left(x + \frac{1}{2}hy, t\right)}{\partial x_j} \right].$$

We fix x, y and t, and consider the function

$$g(z) = \sum_{j=1}^{n} y_j \left[\frac{\partial \Phi\left(x - \frac{1}{2}zy, t\right)}{\partial x_j} - \frac{\partial \Phi\left(x + \frac{1}{2}zy, t\right)}{\partial x_j} \right], \qquad z \geq 0.$$

Since $g(0) = 0$, we have $g(z) = g'(\theta z)z$, $0 < \theta < 1$, and so

$$g(z) = -\sum_{k,j=1}^{n} y_j \left[\frac{\partial \Phi^2\left(x - \frac{1}{2}\theta zy, t\right)}{\partial x_j \partial x_k} + \frac{\partial \Phi^2\left(x + \frac{1}{2}\theta zy, t\right)}{\partial x_j \partial x_k} \right] y_k z. \tag{2.22}$$

Setting $z = h$ we obtain from (2.22)

$$g(h) = -h \sum_{k,j=1}^{n} \left[\frac{\partial \Phi^2\left(x - \frac{1}{2}\theta hy, t\right)}{\partial x_j \partial x_k} y_j y_k + \frac{\partial \Phi^2\left(x + \frac{1}{2}\theta hy, t\right)}{\partial x_j \partial x_k} y_j y_k \right].$$

By condition 6

$$\sum_{k,j=1}^{n} \left[\frac{\partial \Phi^2\left(x - \frac{1}{2}\theta hy, t\right)}{\partial x_j \partial x_k} + \frac{\partial \Phi^2\left(x + \frac{1}{2}\theta hy, t\right)}{\partial x_j \partial x_k} \right] y_j y_k \leq 0.$$

Therefore, $g(h) \geq 0$, which corresponds to (2.21).

Since $F = F_1 - F_2$, we have $F|_{t=a} = 0$, $F|_{t=b} = 0$, $F|_\Gamma = 0$.

Integrating (2.15) over x, p, and t, recalling (2.16)–(2.20) and using the specific properties of the divergence terms and the Cartesian product representation $D \times \mathbf{R}^n \times [a, b]$ of the domain, we obtain the fundamental identity

$$\begin{aligned} &\frac{1}{2} \int_a^b \int_D \int_{\mathbf{R}^n} \sum_{i,j=1}^{n} \left[-\frac{\partial^2 H}{\partial x_i \partial x_j} \frac{\partial F}{\partial p_i} \frac{\partial F}{\partial p_j} + \frac{\partial^2 H}{\partial p_i \partial p_j} \frac{\partial F}{\partial x_i} \frac{\partial F}{\partial x_j} \right] \mathrm{d}p\mathrm{d}x\mathrm{d}t \\ &+ \frac{1}{2h} \int_a^b \int_D \int_{\mathbf{R}^n} \left|\bar{F}\right|^2 \sum_{j=1}^{n} \frac{\partial}{\partial x_j}[\Phi] y_j \mathrm{d}y\mathrm{d}x\mathrm{d}t \\ &+ \int_a^b \int_D \int_{\mathbf{R}^n} (F_1 - F_2)(MF_1 - MF_2)\, \mathrm{d}x\mathrm{d}p\mathrm{d}t = 0. \end{aligned} \tag{2.23}$$

Since

$$\sum_{i,j=1}^{n} \frac{\partial^2 H}{\partial p_i \partial p_j} \frac{\partial F}{\partial x_i} \frac{\partial F}{\partial x_j} \geq \alpha \sum_{i=1}^{n} \left(\frac{\partial F}{\partial x_i} \right)^2, \quad -\sum_{i,j=1}^{n} \frac{\partial^2 H}{\partial x_i \partial x_j} \frac{\partial F}{\partial p_i} \frac{\partial F}{\partial p_j} \geq 0,$$

and (as proved above) $\sum_{j=1}^{n} \frac{\partial}{\partial x_j}[\Phi] y_j \geq 0$ (2.21) implies that

$$\int_a^b \int_D \int_{\mathbf{R}^n} \left\{ \frac{\alpha}{2} \left|\mathrm{grad}_x (F_1 - F_2)\right|^2 + (F_1 - F_2)(MF_1 - MF_2) \right\} \mathrm{d}x\mathrm{d}p\mathrm{d}t \leq 0,$$

which for $F_1 \neq F_2$ contradicts condition 11. Therefore we must have $F_1 = F_2$. From (2.1) it follows in addition that $\lambda_1 = \lambda_2$. The theorem is proved.

Some conditions of the theorem are essential. We give an example of the nonuniqueness for the case when $H(x,p)$ is specified.

Assume that $Q = \{(x,p) : x^2 + p^2 \leq 1\}$, i.e. Q is not a Cartesian product, and $F(x,p) = 1 - \mathrm{e}^{1-(x^2+p^2)}$, $H = \frac{1}{2}(x^2+p^2)$. We have $F|_{\partial Q} = 0$, $\{F, H\} = 0$, $F \neq 0$.

REMARK 2.4. Under the condition

$$\Delta\tilde{\Phi} = 4\pi\sigma \int_{\mathbf{R}^n} F(x,p,t)\mathrm{d}p$$

or with the information of type (2.5) available, it is possible to pose the question of simultaneously determining four functions (F, H, λ, Φ).

2.7 THE EFFECT OF THE 'REDUNDANT' EQUATION

In this section we restrict ourselves to the linear equation of transport because this effect may be present in more general kinetic equations including nonlinear ones. Let us consider the radiation transport equation

$$\frac{1}{v}\frac{\partial F}{\partial t} + \Omega\nabla_{\bar{x}}F + \sigma(\bar{x},t)F = c\sigma\int_B f(\bar{x},t,\Omega\Omega')F\mathrm{d}\Omega' + \frac{1}{v}q,$$

where $\bar{x} = (x,y,z) \in D \subset \mathbf{R}^3$, $t \in \mathbf{R}^1$, $|\Omega| = 1$, D is a domain, B is a unit sphere centered at the origin of coordinates. Expanding the solution $F(\bar{x},\Omega,t)$, the indicatrix $f(\bar{x},\Omega,t)$ and the source function $q(\bar{x},\Omega,t)$ in terms of spherical harmonics gives

$$Y_{lm}(\Omega) = \left[\frac{2l+1}{4\pi}\frac{(l-|m|)!}{(l+|m|)!}\right]^{1/2}(-1)^{1/2(m+|m|)}P_l^{|m|}(\cos\theta)\,\mathrm{e}^{\mathrm{i}m}\varphi,$$

$$\Omega = (\cos\varphi\sin\theta, \sin\varphi\sin\theta, \cos\theta), \quad 0 \leq \varphi \leq 2\pi, \quad -\frac{1}{2}\pi \leq \theta \leq \frac{1}{2}\pi.$$

We can obtain an infinite system of equations for the expansion coefficients $\psi_{lm}(\bar{x},t)$ of the solution $F(\bar{x},p,t)$, see (Case and Zweifel, 1967). This system of equation runs as

$$\begin{aligned}\frac{1}{v}\frac{\partial\psi_{lm}}{\partial t} + \frac{1}{2l+1}\Bigg[&(l+1+m)^{1/2}(l+1-m)^{1/2}\frac{\partial\psi_{l+1,m}}{\partial z} + (l-m)^{1/2}(l+m)^{1/2}\frac{\partial\psi_{l-1,m}}{\partial z}\\
&-\frac{1}{2}\left(\frac{\partial}{\partial x} - \mathrm{i}\frac{\partial}{\partial y}\right)\Big\{(l+m)^{1/2}(l+m-1)^{1/2}\psi_{l-1,m-1}\\
&\qquad -(l-m+2)^{1/2}(l-m+1)^{1/2}\psi_{l+1,m-1}\Big\}\\
&-\frac{1}{2}\left(\frac{\partial}{\partial x} + \mathrm{i}\frac{\partial}{\partial y}\right)\Big\{-(l-m)^{1/2}(l-m-1)^{1/2}\psi_{l-1,m+1}\\
&\qquad +(l+m+1)^{1/2}(l+m+2)^{1/2}\psi_{l+1,m+1}\Big\}\Bigg]\end{aligned}$$

$$= \psi_{lm}\sigma(cf_i - 1) + \frac{1}{v}Q_{lm}, \quad | m | \leq l \tag{2.24}$$

Here $f_l(\bar{x},t)$ are the expansion coefficients of the scattering indicatrix $f(\bar{x},t,\Omega\Omega')$, Q_{lm} are the expansion coefficients of the source function $q(\bar{x},\Omega,t)$.

With respect to the inverse problem, the system of equations (2.24) has a very interesting property: it remains a closed system of evolution equations with respect to variable z while neglecting an infinite set of relations corresponding to the subscripts $l = | m |$. This property depends only on the term $\Omega\nabla_x F$ and allows us to determine, apart from the solution $F(\bar{x},\Omega,t)$, various coefficients of the transport equation, if the proper boundary information is available, e.g. $F_0 = F|_{z=0}$. Furthermore, in the case of finite expansions the elimination of the sections and the scattering indicatrix from (2.24) leads to a sequence of linear problems. We shall restrict our consideration only to these systems of equations.

Let

$$\psi_{lm} = \alpha_{lm} + i\beta_{lm}, \quad Q_{lm} = p_{lm} + iq_{lm}, \quad 0 \leq l \leq N, \quad 0 \leq m \leq l,$$

$$\psi_{N+jm} = 0, \quad Q_{N+jm} = 0, \quad j \geq 1, \quad \beta_{l0} = 0.$$

Theorem 2.11. *The closed system of equations (2.24) with respect to the real functions α_{lm}, β_{lm} which does not contain the unknown coefficients $\sigma(cf_i - 1)$, $l = 0,1,2,\ldots,N$, is as follows:*

$$\frac{\partial \alpha_{Nm}}{\partial z} = \frac{1}{2}\left(\frac{N+m}{N-1+m}\right)^{1/2}\left(\frac{\partial \alpha_{Nm-1}}{\partial x} + \frac{\partial \beta_{Nm-1}}{\partial y}\right)$$

$$-\frac{1}{2}\left(\frac{N-m}{N+1+m}\right)^{1/2}\left(\frac{\partial \alpha_{Nm+1}}{\partial x} + \frac{\partial \beta_{Nm+1}}{\partial y}\right),$$

$$\frac{\partial \beta_{Nn}}{\partial z} = \frac{1}{2}\left(\frac{N+n}{N-1+n}\right)^{1/2}\left(\frac{\partial \beta_{Nn-1}}{\partial x} + \frac{\partial \alpha_{Nn-1}}{\partial y}\right)$$

$$-\frac{1}{2}\left(\frac{N-n}{N+1+n}\right)^{1/2}\left(\frac{\partial \beta_{Nn+1}}{\partial x} + \frac{\partial \alpha_{Nn+1}}{\partial y}\right),$$

$$m = 0,1,\ldots,N, \qquad n = 1,2,\ldots,N$$

$$\alpha_{ll}\frac{\partial \alpha_{l-1,m}}{\partial z} = \frac{\alpha_{ll}}{(l-m)^{1/2}(l+m)^{1/2}}\times$$

$$\left[-\frac{2l+1}{v}\frac{\partial \alpha_{lm}}{\partial t} + \frac{2l+1}{v}p_{lm} - (l+1-m)^{1/2}(l+m+1)^{1/2}\frac{\partial \alpha_{l+1,m}}{\partial z}\right.$$

$$+\frac{1}{2}(l+m)^{1/2}(l+m-1)^{1/2}\left(\frac{\partial \alpha_{l-1,m-1}}{\partial x} + \frac{\partial \beta_{l-1,m-1}}{\partial y}\right)$$

$$-\frac{1}{2}(l-m)^{1/2}(l-m-1)^{1/2}\left(\frac{\partial \alpha_{l-1,m+1}}{\partial x} + \frac{\partial \beta_{l-1,m+1}}{\partial y}\right)$$

$$-\frac{1}{2}(l-m+2)^{1/2}(l-m+1)^{1/2}\left(\frac{\partial \alpha_{l+1,m-1}}{\partial x} + \frac{\partial \beta_{l+1,m-1}}{\partial y}\right)$$

$$-\frac{1}{2}(l+m+1)^{1/2}(l+m+2)^{1/2}\left(\frac{\partial \alpha_{l+1,m+1}}{\partial x} + \frac{\partial \beta_{l+1,m+1}}{\partial y}\right)$$

$$+\frac{\alpha_{lm}}{(l-m)^{1/2}(l+m)^{1/2}}\left[\frac{2l+1}{v}\frac{\partial \alpha_{ll}}{\partial t} - \frac{2l+1}{v}p_{ll} + (2l+1)^{1/2}\frac{\partial \alpha_{l+1,l}}{\partial z}\right.$$

$$-\frac{(2l)^{1/2}(2l-1)^{1/2}}{2}\left(\frac{\partial\alpha_{l-1,l-1}}{\partial x}+\frac{\partial\beta_{l-1,l-1}}{\partial y}\right)$$

$$+\frac{\sqrt{2}}{2}\left(\frac{\partial\alpha_{l+1,l-1}}{\partial x}+\frac{\partial\beta_{l+1,l-1}}{\partial y}\right)$$

$$-\frac{(2l+1)^{1/2}(2l+2)^{1/2}}{2}\left(\frac{\partial\alpha_{l+1,l+1}}{\partial x}+\frac{\partial\beta_{l+1,l+1}}{\partial y}\right)\Bigg],$$

$$\beta_{ll}\frac{\partial\beta_{l-1,n}}{\partial z}=\frac{\beta_{ll}}{(l-n)^{1/2}(l+n)^{1/2}}\times$$

$$\Bigg[-\frac{2l+1}{v}\frac{\partial\alpha_{\rm ln}}{\partial t}+\frac{2l+1}{v}q_{\rm ln}-(l+1-n)^{1/2}(l+n+1)^{1/2}\frac{\partial\beta_{l+1,n}}{\partial z}$$

$$+\frac{1}{2}(l+n)^{1/2}(l+n-1)^{1/2}\left(\frac{\partial\beta_{l-1,n-1}}{\partial x}+\frac{\partial\alpha_{l-1,n-1}}{\partial y}\right)$$

$$-\frac{1}{2}(l-n+1)^{1/2}(l-n-2)^{1/2}\left(\frac{\partial\beta_{l+1,n-1}}{\partial x}+\frac{\partial\alpha_{l+1,n-1}}{\partial y}\right)$$

$$-\frac{1}{2}(l-n)^{1/2}(l-n+1)^{1/2}\left(\frac{\partial\beta_{l-1,n+1}}{\partial x}+\frac{\partial\alpha_{l-1,n+1}}{\partial y}\right)$$

$$-\frac{1}{2}(l+n+1)^{1/2}(l+n+2)^{1/2}\left(\frac{\partial\beta_{l+1,n+1}}{\partial x}+\frac{\partial\alpha_{l+1,n+1}}{\partial y}\right)$$

$$+\frac{\beta_{\rm ln}}{(l-n)^{1/2}(l+n)^{1/2}}\Bigg[(2l+1)\frac{\partial\beta_{ll}}{\partial t}-\frac{2l+1}{v}q_{ll}+(2l+1)^{1/2}\frac{\partial\beta_{l+1,l}}{\partial z}$$

$$-\frac{(2l)^{1/2}(2l-1)^{1/2}}{2}\left(\frac{\partial\beta_{l-1,l-1}}{\partial x}+\frac{\partial\alpha_{l-1,l-1}}{\partial y}\right)$$

$$+\frac{\sqrt{2}}{2}\left(\frac{\partial\beta_{l+1,l-1}}{\partial x}+\frac{\partial\alpha_{l+1,l-1}}{\partial y}\right)$$

$$-\frac{(2l+1)^{1/2}(2l+2)^{1/2}}{2}\left(\frac{\partial\beta_{l+1,l+1}}{\partial x}+\frac{\partial\alpha_{l+1,l+1}}{\partial y}\right)\Bigg],$$

$$l=1,\ldots,N.$$

Assume that $\alpha_{\rm ln}$, $\beta_{\rm ln}$ are the solutions to the above system of equations, then the sought coefficients $\sigma(cf_l-1)$, $l=0,1,2,\ldots,N$, may be calculated using $\alpha_{\rm ln}$, $\beta_{\rm ln}$ by the formula:

$$\sigma(cf_l-1)\alpha_{ll}=\frac{\partial\alpha_{ll}}{\partial t}+\frac{1}{2l+1}\Bigg[(2l+1)^{1/2}\frac{\partial\alpha_{l+1,l}}{\partial z}$$

$$-\frac{(2l)^{1/2}(2l-1)^{1/2}}{2}\left(\frac{\partial\alpha_{l-1,l-1}}{\partial x}+\frac{\partial\beta_{l-1,l-1}}{\partial y}\right)$$

$$+\frac{\sqrt{2}}{2}\left(\frac{\partial\alpha_{l+1,l-1}}{\partial x}+\frac{\partial\beta_{l+1,l-1}}{\partial y}\right)$$

$$-\frac{(2l+1)^{1/2}(2l+2)^{1/2}}{2}\left(\frac{\partial\alpha_{l+1,l+1}}{\partial x}+\frac{\partial\beta_{l+1,l+1}}{\partial y}\right)\Bigg]-p_{ll}.$$

As an example, we give the following result.

Theorem 2.12. *Let the coefficients $\alpha_{ll}^0(x,y,t)$, $\beta_{ll}^0(x,y,t)$ in the expansion of the given function $F_0(x,y,t,\Omega) = F|_{z=0}$ be non-zero in a certain neighbourhood of the origin of (x,y,t)-coordinates. Then there exists a neighbourhood of the origin of (x,y,z,t)-coordinates such that the inverse problem of simultaneously determining the function ψ_{lm} and $\sigma(cf_l - 1)$, $|m| \leq l$ has no more than one solution.*

We should emphasize once again that the effect of 'redundant' equations may be present in more general kinetic equations as well, since this effect depends only on the term $v\Omega \operatorname{grad} F$.

2.8 PROBLEM OF SEPARATION

A solution of kinetic equations appears to have another interesting property: under a certain restriction on the source using the kinetic equation solution trace it is possible to represent the distribution function as a sum of two terms, one produced by the initial state and the other generated by the action of the source.

We shall consider a kinetic equation of the form

$$\frac{\partial F}{\partial t} + \bar{v}\operatorname{grad}_x F = \operatorname{St} F + f, \quad t \in \mathbf{R}^1, \quad \bar{v} \in \mathbf{R}^n, \quad x \in \mathbf{R}^n, \tag{2.25}$$

where $\operatorname{St} F$ is the collision integral, and $f(x,\bar{v},t)$ is a certain function. The restriction concerning the right-hand side of (2.25) is as follows: the function $q(x,\bar{v},t) = \operatorname{St} F + f$ is independent of the absolute value $|\bar{v}|$ of the velocity $\bar{v}$. In the linear theory this restriction is definitely satisfied for isotropic scattering or for $\partial f/\partial |\bar{v}| = 0$. In nonlinear cases, for $\partial f/\partial |\bar{v}| = 0$, the function q is independent of $|\bar{v}|$ if the kernels of the operator St are independent of $|\bar{v}|$. The general solution of equation (2.25) can be written as

$$F(x,\bar{v},t) = F_0(x - \bar{v}t, \bar{v}) + \int_0^1 q(x - \bar{v}t(1-\xi), \bar{v}, t\xi)\mathrm{d}\xi.$$

Let $\bar{v} = \alpha\Omega$, $|\Omega| = 1$. Since q is not assumed to depend on α, then the following equality is valid:

$$F(x,\alpha\Omega,t) = F_0(x - \alpha\Omega t, \alpha\Omega) + t\int_0^1 q(x - \alpha\Omega t(1-\xi), \Omega, t\xi)\mathrm{d}\xi. \tag{2.26}$$

Let Ω and x be fixed, and for $\Omega = \Omega_0$ and $x = x_0$

$$a(\alpha,t) = F, \quad b(\alpha,t) = F_0, \quad c(\alpha,t) = \int_0^1 q\mathrm{d}\xi.$$

From (2.26) we obtain the equality

$$a(\alpha,t) = b(\alpha,t) + c(\alpha,t) \tag{2.27}$$

Here the function b is generated by the initial state F_0, and c by the function q with $x = x_0$, $\Omega = \Omega_0$.

Theorem 2.13. *If the functions a, b, c are analytic in the neighbourhood of the origin of coordinates, then the function a in equality (2.27) uniquely determines the functions b and c.*

PROOF. Bearing in mind possible generalizations, it is worthwhile performing the proof. It is clear from (2.26) that the function b depends on the variable combination $(-\alpha t, \alpha)$, while the function c depends on another combination, $(-\alpha t, t)$, and thus we may write down the relation

$$a(\alpha, t) = b(-\alpha t, \alpha) + tc(-\alpha t, t).$$

Here it is clear that in expanding the functions b and c in the Taylor series, the powers α of are greater than or equal to the powers of t for the function b, while, on the other hand, the powers of t are strictly greater than the powers of α for the function c. Therefore, b and c can be uniquely separated out of the expansion of function a, i.e. it is possible to split the distribution function. The theorem is proved.

REMARK 2.5. On splitting the distribution function the inverse problems of determining the function q can be solved by the familiar techniques assuming the appropriate information is available. These problems will belong to the integral geometry and consist of an integration along lines.

Of course, other results achieved in separating a solution are also of interest including those concerning more general kinetic equations. Note also that the quality of being analytic for the functions in Theorem 2.13 is essential: there are proper examples for the opposite case.

2.9 DIFFERENTIAL AND INTEGRO-DIFFERENTIAL IDENTITIES

In studying the uniqueness and existence of solutions of multidimensional inverse problems for kinetic equations the integro-differential relations are of substantial use (Anikonov, 1985; 1986; Anikonov and Amirov, 1983). We shall give some examples of them here and some applications in Chapter 4.

Theorem 2.14. *If $F(x,p,t)$ is a twice differentiable function and*

$$LF = \frac{\partial F}{\partial t} + \{F, H\}, \qquad BF = \sum_{j=1}^{n} \frac{\partial F}{\partial x_j}\frac{\partial F}{\partial p_j},$$

then the identity

$$LBF = \sum_{j=1}^{n} \frac{\partial F}{\partial p_j}\frac{\partial LF}{\partial x_j} + \frac{\partial F}{\partial x_j}\frac{\partial LF}{\partial p_j} + \sum_{i,j=1}^{n} \left(\frac{\partial^2 H}{\partial x_i \partial x_j}\frac{\partial F}{\partial p_i}\frac{\partial F}{\partial p_j} - \frac{\partial^2 H}{\partial p_i \partial p_j}\frac{\partial F}{\partial x_i}\frac{\partial F}{\partial x_j} \right)$$

is valid.

The quantum equations of type (2.1) turn out to be similar relations which generalize the previously known ones. Here we restrict ourselves only to two equalities. Let F_1 and F_2 be two infinitely differentiable solutions of (2.1) with $a = 0$, which rapidly vanish at 0 with respect to the variable p, such that $F_1 = F_2$ for $t = \alpha$, $t = \beta$, $x \in \partial D$.

Let

$$F = F_1 - F_2, \quad MF_k = \sum_{j=1}^{n} \frac{\partial^2}{\partial x_j \partial p_j} \operatorname{St} F_k, \quad k = 1, 2,$$

$$[\Phi] = \left[\Phi\left(x - \frac{1}{2}hy, t\right) - \Phi\left(x + \frac{1}{2}hy, t\right)\right],$$

where $\hat{F}(x,y,t)$ is the Fourier transform of the function $F(x,p,t)$ with respect to p.

Theorem 2.15. *The equalities*

$$\int_a^b \int_D \int_{\mathbf{R}^n} \left\{|\mathrm{grad}_x F|^2 + F\left(MF_1 - MF_2\right)\right\} dxdpdt + \int_a^b \int_D \int_{\mathbf{R}^n} \left|\hat{F}\right|^2 \sum_{j=1}^n y_j \frac{\partial}{\partial x_j}[\Phi] dydxdt = 0,$$

$$\int_a^b \int_D \int_{\mathbf{R}^n} \left\{ |\mathrm{grad}_x F|^2 + F(MF_1 - MF_2) - \sum_{m=0}^{\infty} \frac{h^{2m}}{(2m+1)!} \sum_{|\alpha|=m+1, |\beta|=m+1} D_x^{\alpha+\beta}\Phi D_p^\alpha F D_p^\beta F \right\} dxdpdt = 0$$

hold.

Corollary I. *f* $\mathbf{M}$ *is a monotone operator and the given function* Φ *is convex with respect to the variable* x, *then* $F = 0$, $x \in D$, $p \in \mathbf{R}^n$, $a \le t \le b$.

The following identity containing the curvature tensor holds for the operator $\mathbf{H}$ in integrating over geodesic lines;

$$\mathbf{H}F = \xi^j \overset{h}{\nabla}_j F$$

where

$$\overset{h}{\nabla}_j = \nabla_j - \Gamma^i_{jk}\xi^k \frac{\partial}{\partial \xi^i},$$

∇_j is the operator of covariant differentiation with respect to the metric tensor $\mathbf{g}_{ij}$, Γ^i_{jk} are the Christoffel symbols of the tensor $\mathbf{g}_{ij}$ (see (Pestov, 1986)).

Theorem 2.16. *The identity*

$$\left|\overset{h}{\nabla} F\right|^2 - \mathbf{R}_{ijkl}\xi^i\xi^k\check{\nabla}^j F\check{\nabla}^l F + \check{\nabla}_i\left(\check{\nabla}^j F \cdot \mathbf{H}F\right)$$

$$= \check{\nabla}_i\left[\overset{h}{\nabla}_j F\left(\xi^i\check{\nabla}^j F - \xi^j\check{\nabla}^i F\right)\right] + 2\overset{h}{\nabla}_i\left(F\check{\nabla}^i \mathbf{H}F\right) - 2FQ\mathbf{H}F$$

holds, where $\mathbf{R}_{ijkl}$ *is the curvature tensor,* $\overset{h}{\nabla}_i = \frac{\partial}{\partial \xi^i}$, $\check{\nabla}^i = \mathbf{g}^{ij}\check{\nabla}_j$, $\overset{h}{\nabla}^i = \mathbf{g}^{ij}\overset{h}{\nabla}_j$, $Q = \check{\nabla}^i \overset{h}{\nabla}_i$ *and* F *is a twice differentiable function.*

Theorem 2.17. *If functions* $F(x,y,z)$, $Q(x,y,z)$, $d(x,y)$, $c(x,y)$ *belong to* $C^2(D)$, $D \in \mathbf{R}^3$ *and*

$$\mathbf{M}F = \left(\frac{\partial F}{\partial x} - c\right)\cos\theta + \left(\frac{\partial F}{\partial y} - d\right)\sin\theta,$$

$$\mathbf{M}'F = \left(\frac{\partial F}{\partial x} - c\right)\sin\theta - \left(\frac{\partial F}{\partial y} - d\right)\cos\theta.$$

Then the identity

$$\mathbf{M}'F\frac{\partial}{\partial z}\mathbf{M}F - \mathbf{M}F\frac{\partial}{\partial z}\mathbf{M}'F = -\frac{\partial\theta}{\partial z}\left[\left(\frac{\partial F}{\partial x} - c\right)^2 + \left(\frac{\partial F}{\partial y} - d\right)^2\right]$$

$$+\frac{\partial}{\partial y}\left(\frac{\partial F}{\partial x}\frac{\partial F}{\partial z}\right)-\frac{\partial}{\partial x}\left(\frac{\partial F}{\partial y}\frac{\partial F}{\partial z}\right)-\frac{\partial}{\partial z}\left(c\frac{\partial F}{\partial y}-d\frac{\partial F}{\partial x}\right)$$

holds.

The use of this kind of identities for geometrical problems is exemplified in (Anikonov and Pestov, 1990).

2.10 SOLUTION EXISTENCE PROBLEMS

The results concerning the existence and uniqueness of the solution in three inverse problems for the transport equation are given in this section. The general technique applied here is developed in (Anikonov, 1986; 1987; Anikonov and Bubnov, 1988; 1989) and in works cited there. By way of illustration, we restrict ourselves to a linear problem to determine a single coefficient.

The boundary value problem for the transport equation

$$\frac{\partial u}{\partial t}+(\omega,\nabla_x u)+\sigma(x,\omega)u=\sigma_s\int_{\Omega} g(x,\omega,\omega')u\,d\omega'+h(t,x,\omega)f(x,\omega), \tag{2.28}$$

$$u|_{\Gamma_-}=\varphi, \tag{2.29}$$

is considered, where $t\in\mathbf{R}$, $x\in D\subset\mathbf{R}^n$, D is a strictly convex bounded domain with boundary $\gamma\in C^2$, $\omega'\in\Omega$, Ω is a sphere of unit radius in $\mathbf{R}^n$ and $\Gamma_-=\{(x,t,\omega)\ :\ t\in\mathbf{R}^1$, $x\in\gamma$, $(\omega,n)<0$, n being the unit vector of the outer normal to $\gamma\}$.

Three inverse problems are posed:

1. Find functions $u(x,t,\omega)$ and $\sigma(x,\omega)$.
2. Find functions $u(x,t,\omega)$ and $\sigma_s(x,\omega)$.
3. Find functions $u(x,t,\omega)$ and $f(x,\omega)$, in (2.28) if conditions (2.28) and (2.29) are supplemented by data for $t=0$ and the stipulation that the solution $u(x,t,\omega)$ behave regularly with respect to the variable t on $(-\infty,\infty)$, i.e. by the following conditions:

$$u\,|_{t=0}=u_0,\quad u_0\,|_{\gamma_-}=\varphi\,|_{t=0},\quad \gamma_-=\{(x,\omega):x\in\gamma,\quad (n,\omega)<0\}, \tag{2.30}$$

$$\max_{x,\omega}\int_{-\infty}^{\infty}(1+|\lambda|)^{2p}\left[|\hat{u}|^2+|\nabla_x\hat{u}|^2\right]d\lambda<\infty,\quad p\geq 2, \tag{2.31}$$

where $\hat{u}$ is the Fourier transform of the function $u(x,t,\omega)$ with respect to the variable t.

That the inverse problems are well-posed is guaranteed by the following assertion.

Theorem 2.18. *Let σ, $\sigma_s\in C^1(\overline{D})\times C(\Omega)$, $g\in C^1$ and let the condition*

$$6d^2\left[\left(\|\sigma\|+\|\sigma_s\|_0\left\|\int_{\Omega}|g|\,d\omega'\right\|\right)^2\right.$$

$$+\left(\|\nabla_x\sigma\|_0+\|\nabla_x\sigma_s\|_0\left\|\int\limits_\Omega |g|\,\mathrm{d}\omega'\right\|_0 \|\sigma_s\|_0\left\|\int\limits_\Omega |\nabla_x g|\,\mathrm{d}\omega\right\|\right)^2\Bigg] \le 1$$

be satisfied. Then, for any functions hf and φ *such that*

$$E(\hat\varphi,\hat h f)=\max_{\omega,x\in\gamma_-}\int\limits_{-\infty}^{\infty}(1+|\lambda|)^{2p}\left[|\hat\varphi|^2+|\nabla_x\hat\varphi|^2\right]\mathrm{d}\lambda$$

$$+\max_{x,\omega}\int\limits_{-\infty}^{\infty}(1+|\lambda|)^{2p}\left[\left|\hat h f\right|^2+\left|\nabla_x\hat h f\right|^2\right]\mathrm{d}\lambda<\infty,\qquad p\ge 2$$

there exists a unique solution $u(x,t,\omega)$ *of problems (2.28), (2.29), and*

$$\max\int\limits_{-\infty}^{\infty}(1+|\lambda|)^{2p}\left[|\hat u|^2+|\nabla_x\hat u|^2\right]\mathrm{d}\lambda\le cE(\hat\varphi,\hat h f),\quad p\ge 2,$$

$$\|\Phi(x,\omega)\|_0=\max_{x,\omega}|\Phi(x,\omega)|,\quad d=\mathrm{diam}\,D.$$

Boundary value problems 1–3 posed above turn out to reduce to an investigation of the boundary value problems for integrodifferential equations (see (Anikonov and Bubnov, 1988)). This reduction is formulated in the following lemmas.

Lemma 2.2. *If* $v(x,\lambda,\omega)$ *is a solution of a boundary value problem*

$$-\mathrm{i}\lambda v+(\omega,\nabla_x v)+v\left[F_0+\frac{\mathrm{i}}{u_0}\int\limits_{-\infty}^{\infty}\lambda v\,\mathrm{d}\lambda\right]=\sigma_s\int\limits_\Omega g v\,\mathrm{d}\omega'+f\hat h, \tag{2.32}$$

$$v\,|_{\gamma_-}=\hat\varphi,\quad |u_0|\ge\sigma>0,\quad \lambda\in\mathbf{R}^1, \tag{2.33}$$

of the class

$$\max_{x,\omega}\int\limits_{-\infty}^{\infty}(1+|\lambda|)^{2p}\left[|v|^2+|\nabla_x v|^2\right]\mathrm{d}\lambda<\infty,\quad p\ge 2,$$

then functions $u(x,t,\omega)$ *and* $\sigma(x,\omega)$ *sought in inverse problem 1 are computed from the formulas*

$$\sigma(x,\omega)=\frac{1}{u_0}F_0+\frac{\mathrm{i}}{u_0}\int\limits_{-\infty}^{\infty}\lambda v\,\mathrm{d}\lambda,\quad u=\int\limits_{-\infty}^{\infty}v\,\mathrm{e}^{-\mathrm{i}\lambda t}\mathrm{d}\lambda,$$

$$F_0=\sigma_s(x,\omega)\int\limits_\Omega g u_0\,\mathrm{d}\omega'-h(0,x,\omega)f(x,\omega)-(\omega,\nabla_x u_0).$$

Lemma 2.3. *If* $v(x,\lambda,\omega)$ *is a solution of a boundary value problem*

$$-\mathrm{i}\lambda v+(\omega,\nabla_x v)+\sigma v=\left[F_1-\frac{\mathrm{i}}{\int\limits_\Omega g u_0\,\mathrm{d}\omega'}\int\limits_{-\infty}^{\infty}\lambda v\,\mathrm{d}\lambda\right]\int\limits_\Omega g v\,\mathrm{d}\omega'+f\hat h, \tag{2.34}$$

$$v|_{\gamma_-}=\hat\varphi,\quad \left|\int\limits_\Omega g u_0\,\mathrm{d}\omega'\right|\ge\sigma>0, \tag{2.35}$$

of the class

$$\left\{v : \max_{x,\omega} \int_{-\infty}^{\infty} (1+ \mid \lambda \mid)^{2p} \left[\mid v \mid^2 + \mid \nabla_x v \mid^2\right] \mathrm{d}\lambda < \infty, \quad p \geq 2\right\}$$

then functions $u(x,t,\omega)$ and $\sigma(x,\omega)$ sought in inverse problem 2 are computed from the formulas

$$\sigma_s = F_1 - \frac{\mathrm{i}}{\int_\Omega g u_0 \mathrm{d}\omega'} \int_{-\infty}^{\infty} \lambda v \mathrm{d}\lambda, \quad u = \int_{-\infty}^{\infty} v\, \mathrm{e}^{-\mathrm{i}\lambda t} \mathrm{d}\lambda,$$

$$F_1 = \frac{(\omega, \nabla_x v) + \sigma u_0 - f(x,\omega) h(0,x,\omega)}{\int_\Omega g u_0 \mathrm{d}\omega'}$$

Lemma 2.4. *If $v(x,\lambda,\omega)$ is a solution of a boundary value problem*

$$-\mathrm{i}\lambda v + (\omega, \nabla_x v) + \sigma v$$

$$= \sigma_s \int_\Omega g v_0 \mathrm{d}\omega' + \frac{\hat{h}(\lambda,x,\omega)}{h(0,x,\omega)} \left[-\mathrm{i} \int_{-\infty}^{\infty} \lambda v \mathrm{d}\lambda + (\omega, \nabla_x u_0) + \sigma u_0 - \sigma_s \int_\Omega g u_0 \mathrm{d}\omega'\right], \tag{2.36}$$

$$v\mid_{\gamma_-} = \hat{\varphi}, \quad |h(0,x,\omega)| \geq \delta > 0, \tag{2.37}$$

of the class

$$\left\{v : \max_{x,\omega} \int_{-\infty}^{\infty} (1+ \mid \lambda \mid)^{2p} \left[\mid v \mid^2 + \mid \nabla_x v \mid^2\right] \mathrm{d}\lambda < \infty, \quad p \geq 2\right\},$$

then functions $u(x,t,\omega)$ and $\sigma(x,\omega)$ sought in inverse problem 3 are computed from the formulas

$$f(x,\omega) = \frac{1}{h(0,x,\omega)} \left[(\omega, \nabla_x u_0) + \sigma u_0 - \sigma_s \int_\Omega g u_0 \mathrm{d}\omega' - \int_{-\infty}^{\infty} \lambda v \mathrm{d}\lambda\right],$$

$$u(x,t,\omega) = \int_{-\infty}^{\infty} v\, \mathrm{e}^{-\mathrm{i}\lambda t} \mathrm{d}\lambda.$$

The solvability of problems (2.32)–(2.37) is established in the following theorems where the condition of the smallness of the data has a somewhat cumbersome form.

Theorem 2.19. *Let $\sigma_s \in C^1(\bar{D}) \times C(\Omega)$, $g \in C^1(\bar{D}) \times C(\Omega) \times C(\Omega)$, $u_0 \in C^2(\bar{D}) \times C(\Omega)$, $\mid u_0 \mid \geq \delta$, $h(0,x,\omega) f \in C^2(\bar{D}) \times C(\Omega)$, $E(\hat{\varphi}, \hat{h} f) < \infty$, and $p \geq 2$ and let*

$$24 d^2 \left(\frac{\|F_0\|^2}{\delta^2} + \|\sigma_s\|_0^2 \left\|\int_\Omega \mid g \mid \mathrm{d}\omega'\right\|_0^2\right) + \frac{(12)^3 d^2 R_1^2}{\delta^2 (2p-3)} + \frac{4(12)^2 d^2 R_1^2}{\delta^4 (2p-3)} \|\nabla_x u_0\|_0^2$$

$$+12 d^2 \left\{\frac{\|F_0\| \, \|\nabla_x u_0\|_0}{\delta^2} + \frac{\|\nabla_x F_0\|_0}{\delta} + \|\nabla_x \sigma_s\|_0 \left\|\int_\Omega \mid g \mid \mathrm{d}\omega'\right\|\right.$$

$$+\|\sigma_s\|_0 \left\| \int_\Omega |\nabla_x g| \, d\omega' \right\| \Bigg\}^2 \le 1,$$

$$R_1 = \max_{\omega, x \in \gamma_-} \int_{-\infty}^{\infty} (1+|\lambda|)^{2p} \left[|\hat{\varphi}|^2 + |\nabla_x \hat{\varphi}|^2 \right] d\lambda + \max_{x,\omega} \int_{-\infty}^{\infty} (1+|\lambda|)^{2p} \left[\left|\hat{h}f\right|^2 + \left|\nabla_x \hat{h}f\right|^2 \right] d\lambda,$$

$$p \ge 2.$$

Then there exists a unique solution of problems (2.32), (2.33) and

$$\max \int_{-\infty}^{\infty} (1+|\lambda|)^{2p} \left[|\hat{v}|^2 + |\nabla_x \hat{v}|^2 \right] d\lambda \le cE(\hat{\varphi}, \hat{h}f).$$

Theorem 2.20. *Let* $\sigma \in C^1(\bar{D}) \times C(\Omega)$, $g \in C^1(\bar{D}) \times C(\Omega) \times C(\Omega)$, $u_0 \in C^2(\bar{D}) \times C(\Omega)$, $\left| \int_\Omega g u_0 d\omega' \right| \ge \delta > 0$, $h(0, x, \omega) f \in C^1(\bar{D}) \times C(\Omega)$, *and* $E(\hat{\varphi}, \hat{h}f) < \infty$. *Let* $\hat{\varphi}$ *and* $\hat{h}f$ *be real function and let*

$$12d^2 \left(\|\sigma\|_0 + \|F_1\|_0 \left\| \int_\Omega |g| \, d\omega' \right\|_0 \right)^2 + \frac{(12)^3 d^2 R_1^2}{\delta^2 (2p-3)} \left\| \int_\Omega |g| \, d\omega' \right\|_0^2$$

$$+12d^2 \left\{ \|\nabla_x \sigma\|_0 + \|\nabla_x F_1\|_0 \left\| \int_\Omega |g| \, d\omega' \right\|_0 + \|F_1\|_0 \left\| \int_\Omega |\nabla_x g| \, d\omega' \right\|_0 \right\}^2$$

$$+\frac{2(12)^2 d^2 R_1^2}{\delta^4 (2p-3)} \left\| \int_\Omega |g| \, d\omega' \right\|_0^2 \left\| \int_\Omega |\nabla_x g u_0| \, d\omega' \right\|_0^2 + \frac{(12)^3 d^2 R_1^2}{\delta^2 (2p-3)} \left\| \int_\Omega |\nabla_x g| \, d\omega' \right\|_0^2 \le 1,$$

$$p \ge 2.$$

Then there exists a unique solution of problems (2.34), (2.35) and

$$\max \int_{-\infty}^{\infty} (1+|\lambda|)^{2p} \left[|\hat{v}|^2 + |\nabla_x \hat{v}|^2 \right] d\lambda \le cE(\hat{\varphi}, \hat{h}f).$$

Theorem 2.21. *Let* $\sigma, \sigma_s \in C^1(\bar{D}) \times C(\Omega)$, $g \in C^1(\bar{D}) \times C(\Omega) \times C(\Omega)$, $u_0 \in C^2(\bar{D}) \times C(\Omega)$, $h(0, x, \omega) \neq 0$, $E(\hat{\varphi}, \hat{h}f) < \infty$ *and* $p \ge 2$. *Let*

$$6d^2 \left[\|\sigma\|_0 + \|\sigma_s\|_0 \left\| \int_\Omega |g| \, d\omega' \right\|_0 + \frac{1}{h(0,x,\omega)} \left(\frac{\int_{-\infty}^{\infty} (1+|\lambda|)^{2p} \left|\hat{h}\right|^2 d\lambda}{2p-3} \right)^{1/2} \right]^2$$

$$+ \left(\|\nabla_x \sigma\|_0 + \|\nabla_x \sigma_s\|_0 \left\| \int_\Omega |g| \, d\omega' \right\|_0 + \|\sigma_s\|_0 \left\| \int_\Omega |\nabla_x g| \, d\omega' \right\|_0 \right)^2 \le 1, \quad p \ge 2.$$

Then there exists a unique solution of problem (2.36), (2.37) and

$$\max \int_{-\infty}^{\infty} (1+\mid \lambda \mid)^{2p} \left[|\hat{v}|^2 + |\nabla_x \hat{v}|^2\right] \mathrm{d}\lambda \leq cE(\hat{\varphi}, \hat{h}f), \quad p \geq 2.$$

The proof of these theorem is based on the study of nonlinear integral equations equivalent to the corresponding boundary value problem. Note that the conditions of Theorems 2.18–2.21 are satisfied with a sufficiently small domain D or with input data sufficiently small in norm.

2.11 AN INVERSE PROBLEM OF MATHEMATICAL BIOLOGY

At present time the theory of differential equations is used for the quantitative and qualitative description of biological processes, for example, the morphogenesis or evolution of population (see (Prigogine, 1980; Murray, 1977; Svirizhev and Logofet, 1978; Pilant and Rundell, 1986; 3Volterra, 1931; Kolmogorov, 1985; Kolmogorov et al., 1985; Lecture Notes in Biomathematics, Springer Verlag, Berlin (1980); Keyfitz, 1968; Rundell, 1989)). Here, often evolution equations of the form

$$\frac{\partial u}{\partial t} = \mathbf{A}u + F(x,t,u), \tag{2.38}$$

arize, where the vector $u(x,t)$ characterizes quantitatively the considered biological substance depending on time t and a classification variable x, and the operator $\mathbf{A}$ and the vector F determine the propagation and interactions of elements of the substance, respectively. Pertaining to type (2.38) are the Lotka–Volterra systems of equations in the theory of evolution of populations, the Turing system of equations describing chemical processes in cells, and the Kolmogorov–Petrovskii–Piskunov equation in the theory of propagation of genes.

For equations of type (2.38) usually initial boundary-value problems are considered and questions of a qualitative aspect with the present biological interpretation.

Presently, inverse problems of mathematical biology are discussed, consisting in defining not only solutions of the considered equations , but also certain coefficients entering into these equations (see (Rundell, 1989)). At the present time new statements are made and systematic studies are undertaken of inverse problems of mathematical biology, where, besides the solution, the sources, as well as the coefficients of diffusion, transport, absorption, birth, death, etc. are described. Here, as in the inverse problems of mathematical physics, the information must be the biological measurements and constraints connected with the initial boundary value data and qualitative behaviour of the solutions.

In this section an inverse problem of determining the absorption coefficient is studied in the equation which describes the evolution of populations, taking into account memory: find two functions $u(x,v,t)$, $\lambda(x,t)$ in the domain $t \in \mathbf{R}$, $v \in \mathbf{R}$, $x \geq 0$ such that the function $\lambda(x,t)$ does not depend on v. It can be a generalized function of the variable, and

$$\frac{\partial u}{\partial t} + \frac{\partial u}{\partial x}v - \frac{\partial \Phi}{\partial x}\frac{\partial u}{\partial v} + \int_{-\infty}^{\infty} \lambda(x,t-\tau)u(x,v,\tau)\mathrm{d}\tau = 0, \tag{2.39}$$

$$u \mid_{x=x_0} = a(v,t), \tag{2.40}$$

where $\Phi(x) > 0$, $x \neq x_0$, $\Phi(x_0) = 0$, $a(v,t)$ are given function, and x_0 is a fixed number.

The biological interpretation is the following: t is time, x is the physiological growth of the specimen, v is the velocity of its change, $u(x,v,t)$ is the number of specimens, $\lambda(x,t)$ is a coefficient characterizing birth and death, the presence of the convolution in (2.38) characterizes the consideration of memory (Kolmogorov, 1985), $\Phi(x)$ is the potential.

Without accounting for physiological growth and for $\Phi(x) = 0$, $v = 1$ the inverse problems for the equation $u_t + u_x + \lambda u$, considered in (Rundell, 1989). Let us also observe that in (Rundell, 1989) the desired function $\lambda(x,t)$ can depend on the solution u, for example through the formula $\lambda = F(x, t, \int_{\mathbf{R}} u dv)$, where F is some function.

EXAMPLE. Let the given function $a(v,t)$ in the inverse problems (2.39), (2.40) be represented in the form

$$a(v,t) = \frac{1}{\sqrt{f(v^2) + \int_{-v}^{v} g\left(\frac{1}{2}v^2 - \frac{1}{2}\eta^2\right) d\eta}} \exp\left[-\frac{1}{4}t^2\left(f(v^2) + \int_{-v}^{v} g\left(\frac{1}{2}v^2 - \frac{1}{2}\eta^2\right) dh\right)^{-1}\right],$$

where $f(y) > 0$, $y \geq 0$, $g(y) \geq 0$ are differentiable functions. Then the functions $u(x,v,t)$, $\lambda(x,t)$ determined by the formulas

$$u(x,v,t) = \frac{1}{\sqrt{f(v^2 + 2\Phi(x)) + \int_{-v}^{v} g\left(\frac{1}{2}v^2 - \frac{1}{2}\eta^2 + \Phi\right) d\eta}}$$

$$\times \exp\left[-\frac{1}{4}t^2\left(f(v^2 + \Phi(x)) + \int_{-v}^{v} g\left(\frac{1}{2}v^2 - \frac{1}{2}\eta^2 + \Phi\right) dh\right)^{-1}\right],$$

$$\lambda(x,t) = -\frac{1}{2}\delta'(t) + g(\Phi(x))\frac{\partial \Phi}{\partial x}\delta''(t),$$

satisfy (2.39), (2.40). Observe that the Fourier transform of the functions u, λ have the form

$$w = \exp\left\{-\left[f(v^2 + 2\Phi(x)) + \int_{-v}^{v} g\left(\frac{1}{2}v^2 - \frac{1}{2}\eta^2 + \Phi(x)\right) d\eta\right]\omega^2\right\},$$

$$\hat{\lambda} = -i\omega - 2g(\Phi(x))\frac{\partial \Phi}{\partial x}\omega^2.$$

Let us consider at first the inverse problems (2.39), (2.40) in the spectral statement: find functions $w(x,v,\omega)$, $\hat{\lambda}(x,\omega)$ in the domain $x \geq 0$, $v \in \mathbf{R}$, $\omega \in \mathbf{R}$ such that

$$\frac{\partial w}{\partial x}v - \frac{\partial w}{\partial v}\frac{\partial \Phi}{\partial x} + \left(\hat{\lambda}(x,\omega) + i\omega\right) = 0, \tag{2.41}$$

$$w\,|_{x=x_0} = \hat{a}(v,\omega). \tag{2.42}$$

Theorem 2.22. *Let the function $\hat{a}(v,\omega)$ be greater than zero and twice continuously differentiable with respect to the variable v, $\hat{a}'_v(0,\omega) = 0$. Then*

$$w(x,v,\omega) = A(v^2 + 2\Phi(x), \omega) \exp\left[\int_{-v}^{v} B\left(\frac{1}{2}v^2 - \frac{1}{2}\eta^2 + \Phi(x), \omega\right) d\eta\right], \tag{2.43}$$

$$\hat{\lambda}(x,\omega) = 2B(\Phi(x),\omega)\frac{\partial \Phi}{\partial x} - \mathrm{i}\omega, \tag{2.44}$$

$$A(y,\omega) = \sqrt{\hat{a}(\sqrt{y},\omega)\hat{a}(-\sqrt{y},\omega)}, \quad y \geq 0, \tag{2.45}$$

$$B(y,\omega) = \frac{1}{2\pi}\frac{\mathrm{d}}{\mathrm{d}y}\int_0^y \ln\sqrt{\frac{\hat{a}(\sqrt{2\eta},\omega)}{\hat{a}(-\sqrt{2\eta},\omega)}}\frac{\mathrm{d}\eta}{\sqrt{y-\eta}}. \tag{2.46}$$

Proof. For the proof of the theorem it is necessary to establish that the functions w, $\hat{\lambda}$ defined by (2.43), (2.44) satisfy (2.41) and (2.42). Let us note that since the given function $\hat{a}, \hat{a}'_v(0,\omega) = 0$ is twice continuously differentiable with respect to v, the functions $A(y,\omega)$, $B(y,\omega)$ for each fixed ω are at least once differentiable with respect to y. We have the relations

$$\frac{\partial w}{\partial x} = \Big[2A'_y(v^2 + 2\Phi(x),\omega)\exp\Big\{\int_{-v}^{v} B\Big(\frac{1}{2}v^2 - \frac{1}{2}\eta^2 + \Phi(x),\omega\Big)\mathrm{d}\eta\Big\}$$

$$+A(v^2 + 2\Phi(x),\omega)\exp\Big\{\int_{-v}^{v} B\Big(\frac{1}{2}v^2 - \frac{1}{2}\eta^2 + \Phi(x),\omega\Big)\mathrm{d}\eta\Big\}$$

$$\times\int_{-v}^{v} B'_y\Big(\frac{1}{2}v^2 - \frac{1}{2}\eta^2 + \Phi(x),\omega\Big)\mathrm{d}\eta\Big]\frac{\partial \Phi}{\partial x}, \tag{2.47}$$

$$\frac{\partial w}{\partial v} = \Big[2A'_y\left(v^2 + 2\Phi(x),\omega\right)\exp\Big\{\int_{-v}^{v} B\Big(\frac{1}{2}v^2 - \frac{1}{2}\eta^2 + \Phi(x),\omega\Big)\mathrm{d}\eta\Big\}$$

$$+A\left(v^2 + 2\Phi(x),\omega\right)\exp\Big\{\int_{-v}^{v} B\Big(\frac{1}{2}v^2 - \frac{1}{2}\eta^2 + \Phi(x),\omega\Big)\mathrm{d}\eta\Big\}$$

$$\times\int_{-v}^{v} B'_y\Big(\frac{1}{2}v^2 - \frac{1}{2}\eta^2 + \Phi(x),\omega\Big)\mathrm{d}\eta\Big]v$$

$$+2A\left(v^2 + 2\Phi(x),\omega\right)B(\Phi,\omega)\exp\Big\{\int_{-v}^{v} B\Big(\frac{1}{2}v^2 - \frac{1}{2}\eta^2 + \Phi(x),\omega\Big)\mathrm{d}\eta\Big\}. \tag{2.48}$$

Multiplying (2.47) by v, (2.48) by $(-\partial\Phi/\partial x)$ and adding the expressions obtained, we arrive at the equation

$$\frac{\partial w}{\partial x}v - \frac{\partial w}{\partial v}\frac{\partial \Phi}{\partial x} = -2B(\Phi,\omega)\frac{\partial \Phi}{\partial x}A(v^2 + 2\Phi,\omega)\exp\Big\{\int_{-v}^{v} B\Big(\frac{1}{2}v^2 - \frac{1}{2}\eta^2 + \Phi(x),\omega\Big)\mathrm{d}\eta\Big\},$$

which can be rewritten in the form

$$\frac{\partial w}{\partial x}v - \frac{\partial w}{\partial v}\frac{\partial \Phi}{\partial x} + \left(\hat{\lambda} + \mathrm{i}\omega\right) = 0,$$

which in fact completes the proof of the first part of the theorem.

Now let us show that $w\,|_{x=x_0} = \hat{a}(v,\omega)$. By hypothesis $\Phi(x_0) = 0$. Therefore, from (2.42) we have

$$w\,|_{x=x_0} = A(v^2,\omega)\exp\left\{\int_{-v}^{v} B\left(\frac{1}{2}v^2 - \frac{1}{2}\eta^2,\omega\right) d\eta\right\}. \tag{2.49}$$

At first let $v \geq 0$. From (2.49) we obtain the equation

$$w\,|_{x=x_0} = A(v^2,\omega)\exp\left\{2\int_{0}^{v} B\left(\frac{1}{2}v^2 - \frac{1}{2}\eta^2,\omega\right) d\eta\right\}.$$

Setting $\frac{1}{2}v^2 - \frac{1}{2}\eta^2 = y$ we find the relation

$$w\,|_{x=x_0} = A(v^2,\omega)\exp\left\{2\sqrt{2}\int_{0}^{\frac{1}{2}v^2} \left(B(y,\omega)\Big/\sqrt{\frac{1}{2}v^2 - y}\right) dy\right\}.$$

Since

$$A = \sqrt{\hat{a}(\sqrt{y},\omega)\hat{a}(-\sqrt{y},\omega)}, \quad B = \frac{1}{2\pi}\frac{\mathrm{d}}{\mathrm{d}y}\int_0^y \ln\sqrt{\frac{\hat{a}(\sqrt{2\eta},\omega)}{\hat{a}(-\sqrt{2\eta},\omega)}}\frac{\mathrm{d}\eta}{\sqrt{y-\eta}},$$

then by Abel's formula of inversion of an integral equation

$$w\,|_{x=x_0} = \sqrt{\hat{a}(v,\omega)\hat{a}(-v,\omega)}\exp\left\{\ln\sqrt{\frac{\hat{a}(v,\omega)}{\hat{a}(-v,\omega)}}\right\} = \hat{a}(v,\omega).$$

Consequently, for $v \geq 0$ the equation $w\,|_{x=x_0} = \hat{a}(v,\omega)$ is established.

Let us prove the equality $w\,|_{x=x_0} = \hat{a}(v,\omega)$ for negative v. From (2.49) we have the relation

$$w(x_0,-v,\omega) = A(v^2,\omega)\exp\left\{-2\int_{0}^{\frac{1}{2}v^2} B\left(\frac{1}{2}v^2 - \frac{1}{2}\eta^2,\omega\right) d\eta\right\}, \quad v \geq 0.$$

Repeating the previous argument, we obtain the required equation

$$w(x_0,-v,\omega) = \hat{a}(-v,\omega), \quad v \geq 0.$$

The theorem is proved.

Let us return to the inverse problem (2.39), (2.40). Besides the restrictions presented above on the variable v for the existence of the Fourier transforms of the functions w, and $\hat{\lambda}$, restrictions are necessary with respect to the variable ω of the information $\hat{a}(v,\omega)$,

$$\hat{a}(v,\omega) = \frac{1}{2\pi}\int_{-\infty}^{\infty} a(v,t)\,\mathrm{e}^{-it\omega}\mathrm{d}t.$$

From the representation of the given function $\hat{a}$ in terms of the function $A(y,\omega)$, $B(y,\omega)$ in the form

$$\hat{a} = A\left(v^2,\omega\right)\exp\left\{\int_{-v}^{v} B\left(\frac{1}{2}v^2 - \frac{1}{2}\eta^2,\omega\right) d\eta\right\}, \tag{2.50}$$

and formulas (2.44), (2.45) it is clear that there is a unique correspondence between the functions $\hat{a} > 0$ and the functions $A(y,\omega)$, and $B(y,\omega)$, $y \geq 0$, characterizing the even and odd parts of the functions $\hat{a}$ with respect to v. We will, in correspondence with this remark, assume that the information $\hat{a}$ is represented in the form (2.50), where $A(y,\omega)$, and $B(y,\omega)$ satisfy the following conditions: A, B are continuously differentiable with respect to y, the function A along with its derivative with respect to y is rapidly decreasing in ω, and the function B along with its derivative with respect to y is bounded. Under these restrictions we have formulas for the functions $u(x,v,t)$, $\lambda(x,t)$ in the inverse problems (2.39), (2.40):

$$u = \int_{-\infty}^{\infty} \left[A(v^2 + 2\Phi(x), \omega) \exp \left\{ \int_{-v}^{v} B\left(\frac{1}{2}v^2 - \frac{1}{2}\eta^2 + \Phi(x), \omega\right) \mathrm{d}\eta \right\} \right] \mathrm{e}^{\mathrm{i}\omega t} \mathrm{d}\omega,$$

$$\lambda(x,t) = \int_{-\infty}^{\infty} \left[2B(\Phi(x), \omega) \frac{\partial \Phi}{\partial x} - \mathrm{i}\omega \right] \mathrm{e}^{\mathrm{i}\omega t} \mathrm{d}\omega.$$

CHAPTER 3

Geometry of Convex Surfaces in the Large and Inverse Problems of Scattering Theory

3.1 GEOMETRICAL QUESTION OF SCATTERING THEORY

In applications we often meet situations, where a source of a physical field or the totality of a physical field is a certain manifold. Physical fields can be electromagnetic, acoustic, seismic, thermal, gravitational, etc. It appears natural to name such manifold-emitting sources manifolds. Earthquake sources, stars are examples of emitting objects. The capability of emission may have various causes: it may be an internal property of the emitting object, or the manifold becomes emitting after external action, or both cases are taking place at the same time. The main problem in this case consists in the reconstruction of the emitting manifold by the field, which it creates. This chapter is devoted to these questions. Our main attention is devoted to the inverse problem of scattering theory, that is the reconstruction of the scatterer by the scattering information. Nevertheless, many results, we believe, may be useful in other situations.

The physical formulation of a variety of inverse problems consists of the following: there is a certain body-scatterer in space, which is irradiated by an electromagnetic, acoustic or another field.

The body scatters the field: the form of the scatterer can be found by scattering information. In physical problems a body often is irradiated from different directions at one or several frequencies in such a way that extensive information on the scatterer is obtained from scattering. The amplitude of the scattered field is a physically authentic information. It is very essential from a mathematical viewpoint, how and where is the field measured, is the source monochromatic or not, what is the nature of the direct wave, etc. Since scattering problems are very complex, asymptotic methods are widely used at investigations and applications. Frequencies, and the distance from the scatterer of the measurement point are usually parameters, by which the asymptotic expansion is produced. In this case geometrical characteristics of the scatterer such as the square of the orthogonal projection on a plane, Gaussian curvature of boundary, characteristic functions, support and other functions arise in the asymptotic expansion as coefficients. This geometrical information is very important, because in certain cases it permits effectively the determination of the form of the scattering body, i.e. to solve the inverse scattering problem. In this connection it is especially essential that methods of geometry in the whole and methods of integral geometry may be used with success. They lead to new, and sometimes exhaustive results. It needs to be underlined once again that geometrical methods operate asymptotic. Therefore theorems of existence, uniqueness and especially solution stability of inverse scattering problem are important here.

Consider a first example, showing the geometrical characteristics in a scattering problem. Many problems of wave propagation lead to the boundary problem for Helmholtz equation. So if a plane wave $u_0(x,p,\lambda) = \exp(\mathrm{i}\lambda < x,p >)$ (λ being the frequency), spreading in the direction p, $|p| = 1$, falls onto scatterer K, bordered by a closed surface B, then for the scattering field $u(x,p,\lambda)$ we have the external problem

$$\Delta u + \lambda^2 u = 0, \quad x \in \mathbf{R}^m \setminus \mathrm{K},$$

$$\Gamma(u(x,p,\lambda) + \exp(\mathrm{i}l < x,p >)|_B = 0,$$

$$\underset{x\to\infty}{u} = \mathrm{O}\left(|x|\right)^{(1-m)/2}, \quad \frac{\partial u}{\partial |x|} - \mathrm{i}\lambda u_{|x|\to\infty} = \mathrm{o}\left(|x|\right)^{(1-m)/2},$$

where Γ is the boundary operator of Dirichlet, Neumann or impedance condition. At $m = 3$ an exact solution of the direct problem (i.e. finding the scattering field $u(x,p,\lambda)$ from known boundary B of scatterer K) is given by the Kirchhoff formula

$$u(x,p,\lambda) = \frac{1}{4\pi}\int\limits_B \left\{u(y,p,\lambda)\frac{\partial}{\partial n_y}\frac{\mathrm{e}^{\mathrm{i}\lambda|x-y|}}{|x-y|} - \frac{\partial u(y,p,\lambda)}{\partial n_y}\frac{\mathrm{e}^{\mathrm{i}\lambda|x-y|}}{|x-y|}\right\} \mathrm{d}s_y,$$

(n_y being outer normal in point $y \in B$).

In the case of a more general equation the solution $u(x,p,\lambda)$ is computed by the formula

$$u(x,p\lambda) = \frac{1}{4\pi}\int\limits_B \left\{u(y,p,\lambda)\frac{\partial}{\partial n_y}G(x,y,\lambda) - \frac{\partial u(y,p,\lambda)}{\partial n_y}G(x,y,\lambda)\right\} \mathrm{d}s_y, \qquad (3.1)$$

where $G(x,y,\lambda)$ is the fundamental solution. In this case solution $u(x,p,\lambda)$ may be a vector, G may be a matrix. In particular, in the case of the dynamical equation of elasticity theory with constant coefficients elements G_{kj} of matrix $\mathbf{G}$ are given by relations

$$G_{kj} = \frac{1}{2\pi\mu}\delta_{kj}\frac{\mathrm{e}^{ik_2|x-y|}}{|x-y|} - \frac{1}{2\pi\rho\lambda^2}\frac{\partial^2}{\partial x_k \partial x_j}\left\{\frac{\mathrm{e}^{ik_1|x-y|} - \mathrm{e}^{ik_2|x-y|}}{|x-y|}\right\},$$

where $k_1^2 = \dfrac{\rho\lambda^2}{\alpha+2\mu}$, $k_2^2 = \dfrac{\rho\lambda^2}{\mu}$, α and μ being Lame constants.

The inverse scattering problem consists of determining the boundary of a simply connected scatterer K by information on the scattered field $u(x,p,\lambda)$, to be known to a certain set M of the variables x,p,λ, i.e. by function $u(x,p,\lambda)|_M$. In applications usually a known function

$$A(\nu,p,\lambda) = \lim_{|x|\to\infty} |x|\,\mathrm{e}^{-\mathrm{i}\lambda|x|}u(|x|\,\nu,p,\lambda)$$

is considered $|\nu| = 1$, $|p| = 1$, $\lambda > 0$, which is called the scattering amplitude or the intensity of the scattering field $|A|$ in the direction ν of observation. By the Kirchhoff formula we have

$$A = -\frac{1}{4\pi}\int\limits_B \mathrm{e}^{-\mathrm{i}\lambda<\nu,y>}\left[\mathrm{i}\lambda < n_y,\nu > u(y,p,\lambda) - \frac{\partial}{\partial n_y}u(y,p,\lambda)\right] \mathrm{d}s_y.$$

Consider the Dirichlet problem, and the scatterer smooth and strictly convex. Applying Kirchhoff's approach (see (Taylor, 1981)), for the scattering amplitude we obtain the following expression

$$A(\nu,p,\lambda)=\frac{\mathrm{i}\lambda}{4\pi}\int\limits_{B}\left[\mathrm{e}^{\,\mathrm{i}\lambda<p-\nu,y>}\left(<n_y,\nu>+|<n_y,p>|\right)\right]\mathrm{d}s_y.$$

In particular, the scattering amplitude in the forward direction $\nu=p$, is proportional to the area $F(p)$ of the shadow of scatterer K;

$$A(p,p,\lambda)=\frac{\mathrm{i}\lambda}{2\pi}\int\limits_{<p,n_y>\geq 0}<p,n_y>\mathrm{d}s_y=\frac{\mathrm{i}\lambda}{2\pi}F(p).$$

In case of $\nu=-p$ (backscattering) the formula

$$A(-p,p,\lambda)=-\frac{\mathrm{i}\lambda}{2\pi}\int\limits_{<p,n_y>\leq 0}\mathrm{e}^{\,2\mathrm{i}\lambda<p,y>}<p,n_y>\mathrm{d}s_y=\rho(p,\lambda)$$

holds.

Notice that if function $\rho(p,\lambda)$ is known for all directions p and frequencies λ, then computing $\rho(p,\lambda)+\rho^*(-p,\lambda)$ (* denotes conjugation) and applying the Ostrogradskii formula we obtain the relation (Levis, 1969) at $\gamma(y)=1,\ y\in\mathrm{K};\ \ \gamma(y)=0,\ y\notin\mathrm{K}$,

$$\pi\frac{\rho(p,\lambda)\rho^*(-p,\lambda)}{\lambda^2}=\int\limits_{-\infty}^{\infty}\int\limits_{-\infty}^{\infty}\int\limits_{-\infty}^{\infty}\gamma(y)\,\mathrm{e}^{\,2\mathrm{i}\lambda<p,y>}\mathrm{d}y=\Phi(\lambda,p),$$

i.e. $\Phi(\lambda,p)$ is the Fourier transformation of the characteristic function $\gamma(y)$ of scatterer K. In practice measurements of $\rho(p,\lambda)$ are produced in a limited range of frequencies land directions.

Another way to solve the scattering problem uses geometrical optics. In this case the asymptotic expansion of the scattering amplitude at $\lambda\to\infty$ contains a summand, with a Gaussian curvature of border of the smooth strictly convex scatterer as a main term of expansion. So in (Majda and Tayler, 1977a) it has been proved that for $\nu\neq p$,

$$\lim_{\lambda\to\infty}|A^*(-\nu,p,\lambda)|^2=\gamma(\nu,p)K^{-1}(\tilde{y}),$$

where γ is reflection coefficient, $K(\tilde{y})$ the Gaussian curvature of the border of the scatterer in point $\tilde{y}$ with normal $n=(\nu-p)/(|\nu-p|)$.

It is possible to come to geometrical characteristics of the scatterer according to (Majda, 1976a) in the following way: consider the mixed problem for hyperbolic equation

$$\frac{\partial^2 u}{\partial t^2}=\sum_{i,j=1}^{m}\frac{\partial}{\partial x_i}\left[a_{ij}(x)\frac{\partial u}{\partial x_j}\right],\quad x\in\mathrm{K},$$

$$u|_{t=0}=f_1(x),\quad \left.\frac{\partial u}{\partial t}\right|_{t=0}=f_2(x),\quad \Gamma u|_{-\mathrm{K}}=0,$$

where $a_{ij}(x)$ are the elements of a smooth positive symmetrical matrix with $a_{ij}=\delta_{ij}$ for $|x|>\rho$ (domain K is contained in a ball $|x|<\rho$). Let $u(x,t)$ be a solution of this problem. Let

$$K^{\pm}(s,p)=\lim_{t\to\pm\infty}tu_t\left[(s+t)p,\right].$$

Functions $K^{\pm}(s,p)$ characterize the behaviour of the solution along ray $x = (t+s)p$ at large positive and negative time. Consider the scattering operator $\mathbf{S}$, which is defined as mapping $\mathbf{S}$: $K^{-}(s,p) \to K^{+}(s,p)$ and permit representation (Majda and Tayler, 1977a)

$$K^{+}(s,p) = \int\int \mathbf{S}(s-\tilde{s},p,\nu)K^{-}(\tilde{s},\nu)\mathrm{d}\tilde{s}\mathrm{d}\nu,$$

where kernel $\mathbf{S}(s,p,\nu)$, $s \in \mathbf{R}^1$, $|p| = |n| = 1$ is a distribution in $D'(\mathbf{R}^1 \times S^2 \times S^2)$. It turns out that for kernel $\mathbf{S}(s,p,\nu)$ the asymptotical expansion

$$\mathbf{S}(s,-p,p) = 2\pi\left(\sum_j K(y_j)^{-1/2}\right)\delta'(s+2h(p)+c_1\delta(s+2h(p))$$
$$+ \text{(smooth members)},$$

holds for p from a certain open subset (dense everywhere) of unit sphere S^2. In this formula y_j are points on ∂K, in which $< y_j, p >= h(p)$, $h(p)$ is the support function of convex hull of surface ∂K; $K(y_j)$ is the Gaussian curvature in point y_j. Moreover, it turns out that

$$A(p) = \max_{s\in\mathbf{R}^1} \text{sing supp } \mathbf{S}(s,-p,p) = -2h(p),$$

i.e. we can obtain the support function of the convex hull of the scatterer from information about scattering operator.

At last, we provide the result, connected with the convex hull of the scatterer. Let $\Phi(x)$ be a finite function of real variable $x \in \mathbf{R}^m$ and let $f(z)$, $z \in C^m$ be determined by the equality

$$f(z) = \left(\frac{1}{\sqrt{2\pi}}\right)^m \int_{\mathbf{R}} \Phi(x)\,\mathrm{e}^{-\mathrm{i}<x,z>}\mathrm{d}x.$$

The scattering problem leads to a similar integral, while, as noted by, for example, (Levis, 1969), the support of function $\Phi(x)$ plays the role of the scatterer. Therefore, defining $\Phi(x)$ becomes an important problem. By the Plancherel–Polya theorem the support function of a convex hull of support of the function $\Phi(x)$ is equal to

$$h(p) = \sup_{x\in\mathbf{R}^m} \lim_{R\to\infty} \frac{1}{R} \ln|f(x+\mathrm{i}Rp)|.$$

In real situations and when investigating inverse problems frequency λ, as a rule, is fixed and there is only a limited range of directions, for which the radiation intensity is measured. Therefore it appears expedient to pick out principal geometrical information, connected with the radiation intensity, and to put this outlined component into the base of the investigation technique of inverse problems to recover the emitting manifold.

By formula (3.1) analysing of various investigation methods of problems, in particular scattering, leads to the necessity of study of problems connected with the representation of the amplitude of the field in the far zone in the form

$$A(n,\lambda,t) = \int_{B(n,t)} \mathrm{e}^{\mathrm{i}\hat{f}(\lambda,y,n,t)}\hat{g}(\lambda,y,n,t)\mathrm{d}s_y, \tag{3.2}$$

where $B(n,t)$ is the illuminated part of manifold $B(t)$ in the direction of n, $\hat{f}$ and $\hat{g}$ are certain functions, t is time, λ is the frequency, and n is the direction of reception of emission. In the case of a regular convex surface $B(t)$ formula (3.2) may be rewritten in the following way

$$A(n,\lambda,t) = \int\limits_{<n,p>\geq 0} e^{i\hat{f}(\lambda,y(p,t),n,t)}\hat{g}(\lambda,y(p,t),n,t)R_1R_2 d\omega_p. \qquad (3.3)$$

Here $y(p,t)$ is the focal radius of surface $B(t)$ in the point with normal p, $|p| = 1$, $R_1(p,t)$, $R_2(p,t)$ are main curvature radii of surface $B(t)$ as functions of normal p and time t. By solution of the Minkowski problem it is possible to find, that $y = \mathbf{M}z$, where $\mathbf{M}$ is an operator, giving $z = R_1(p,t)R_2(p,t)$ corresponding to point $y(p,t)$ of surface B. That is why (3.3) can be rewritten as

$$A(n,\lambda,t) = \int\limits_{<n,p>\geq 0} e^{i\hat{f}(\lambda,\mathbf{M}z,n,t)}\hat{g}(\lambda,\mathbf{M}z,n,t)z(p,t)d\omega_p. \qquad (3.4)$$

As it was mentioned above, the really measured information is $|A|$. That is why, and by the above in the stationary case when

$$A(n,\lambda) = \int\limits_{<n,p>\geq 0} e^{i\hat{f}(\lambda,\mathbf{M}z,t)}\hat{g}(\lambda,\mathbf{M}z,n)z(p)d\omega_p$$

from which we separate the main geometrical information, connected in with the scattering intensity. Namely, for $|A|$ we use the representation of the solution of the inverse problem

$$|A| = a\int\limits_{\omega} |< n,p >| R_1R_2 d\omega_p + b\int\limits_{\omega} \chi(< n,p >)R_1R_2 d\omega_p + Q. \qquad (3.5)$$

Here ω is a unit sphere in $\mathbf{R}^3$; $\int\limits_{\omega} |< n,p >| R_1R_2 d\omega_p$ is the area of orthogonal projection of the concave surface to plane, orthogonal to n; χ is the Heaviside function; $\int\limits_{\omega} \chi(< n,p >)R_1R_2 d\omega_p$ is the area of the illuminated part of the concave surface in the direction of n; a and b are constants; Q is the rest. Relation (3.5) is taken as the basis for further investigations. In this case if $\mathbf{P}^{-1}$ is the inverse operator

$$\mathbf{P}z = a\int\limits_{\omega} |< n,p >| z(p)d\omega_p + b\int\limits_{\omega} \chi(< n,p >)z(p)d\omega_p,$$

then from (3.5) we obtain an equation of the second kind with respect to the sought function $z(p) = R_1R_2$, which has the form

$$z(p) = \tilde{A}(p) + Tz,$$

here

$$\tilde{A}(p) = \mathbf{P}^{-1}|A|, \quad Tz = \mathbf{P}^{-1}Qz,$$

and to which standard methods of the exploration technique may be applied, founded, for instance, in the theory of fixed points. In this connection principal attention will be given to operator $\mathbf{P}$:

$$\mathbf{P}z = a\int\limits_{\omega} |< n,p >| z(p)d\omega_p + b\int\limits_{\omega} \chi(< n,p >)z(p)d\omega_p.$$

3.2 INTEGRAL EQUATION OF THE FIRST KIND

In this section integral equations of the first kind are considered which are the result of inverse problems of scattering theory. In particular, geometrical aspects to define the convex surface by the functional from its orthogonal projections (shadows) and illuminated parts.

Induce notation is used in this section. Let n, p be points of the unit sphere $\omega = x \in \mathbf{R}^3,\ |x| = 1$. For function $f(n)$, given at ω, let $\tilde{f}(\gamma, n)$ be denoted as its average value on the circumference $\gamma = \text{const}$, when point n is a pole of the spherical coordinate system

$$\tilde{f}(\gamma, n) = \frac{1}{2\pi} \int\limits_0^{2\pi} f(p(\gamma, \tau))\mathrm{d}\tau.$$

Here g, t are spherical coordinates of the point $p \in \omega$ with respect to pole n. As $f^{\pm}$ are the even and odd parts of the function $f(n)$, that is $f^{\perp} = \frac{1}{2}\,[f(n) \pm f(-n)]$, Δ is the Laplace–Beltrami operator on ω, $<>$ means a scalar product, $\omega(n) = p \in \omega,\ < n, p > \geq 0$ is a hemisphere, depending on $n \in \omega$, $\mathbf{p}(n, p) = a\,|< n, p >| + b\chi(< n, p >)$, χ is Heaviside function, a,b are constants.

Consider the first kind equation with respect to measures μ on ω.

$$f(n) = \int\limits_{\omega} \mathbf{P}(n, p)\mu(\mathrm{d}\omega_{\mathbf{p}}), \quad n \in \omega. \tag{3.6}$$

Theorem 3.1. *If $a \neq 0$, $b \neq 0$, $a + b \neq 0$ then equation (3.6) has not more than one solution μ.*

PROOF. Let μ be a solution of equation (3.6). Show, that it is unique. Since at any $p \in \omega$ the function $|< n, p >|$ is even in the variable n, then by (3.6) we obtain equations

$$f^{+}(n) = a \int\limits_{\omega} |< n, p >|\,\mu^{+}(\mathrm{d}\omega_{\mathbf{p}}) + \frac{b}{2} \int\limits_{\omega} \mu^{+}(\mathrm{d}\omega_{\mathbf{p}}) + \frac{b}{2} \int\limits_{\Gamma(n)} \mu^{+}(\mathrm{d}\omega_{\mathbf{p}}), \tag{3.7}$$

$$f^{-}(n) = b \int\limits_{\omega(n)} \mu^{-}(\mathrm{d}\omega_{\mathbf{p}}), \tag{3.8}$$

where $\Gamma(n) = p \in \omega,\ < n, p >= 0$ is circumference, and μ^{+}, μ^{-} are the even and odd part of measure μ, respectively. That is $\mu^{\pm} = \frac{1}{2}\,[\mu(Q) + \mu(-Q)]$ for any set of Q on ω, $-Q$ is the set diametrically opposite to Q. Show that equation (3.7) identifies uniquely with the even part μ^{+} of measure, and equation (3.8) identifies with the odd part μ^{-} of measure μ.

Consider integral

$$\int\limits_{\Gamma(n)} \mu^{+}(\mathrm{d}\omega_{\mathbf{p}}), \quad n \in \omega.$$

Let $\mu = \tilde{\mu} + \mu_{\mathbf{sing}}$ be an Lebesgue expansion of measure μ to the amount of being absolutely continuous and singular with respect to Lebesgue measure m on ω. Then

$$\int\limits_{\Gamma(n)} \mu^{+}(\mathrm{d}\omega_{\mathbf{p}}) = \mu^{+}_{\mathbf{sing}}(\Gamma(n)).$$

Support of the singular measure has zero Lebesgue measure, therefore $\mu^+(\Gamma(n)) = 0$ on the set of zero m-measure. Integrating both parts by sphere ω and noting that

$$\int_\omega |< n, p >| d\omega_n = 2\pi, \quad \int_\omega \mu^+_{\text{sing}}(\Gamma(\bar{n}))d\omega_n = 0,$$

we obtain the equality

$$\int_\omega f^+(n)d\omega_n = 2\pi(a+b)\int_{i\omega} \mu^+(d\omega_p).$$

With this we can rewrite the equation as

$$f_1(n) = a\int_\omega |< n, p >| \mu^+(d\omega_p) + \frac{b}{2}\mu^+_{\text{sing}}(\Gamma(n)). \tag{3.9}$$

From hereon for any function $f(n)$ is designated by $f_1(n)$:

$$f_1(n) = \frac{1}{a}f^+(n) - \frac{b}{4\pi a(a+b)}\int_\omega f^+(n)d\omega_n.$$

Furthermore, by proof of (Aleksandrov, 1937a,b; 1938a,b) and with the assumption that $\mu_{\text{sing}}(\Gamma(n)) = 0$ almost everywhere at ω, we obtain that equation (3.3) uniquely identifies with the even part of the measure μ. For the proof of the uniqueness of the solution of equation (3.8), obviously, it is sufficiently to show, that from the equality

$$\int_{\omega(n)} \mu^-(d\omega_p) = 0, \quad n \in \omega,$$

follows $\mu^- = 0$. Draw two arbitrary plane through a straight line, passing through the origin and belonging to a plane. Let these be planes $P(n_1)$ and $P(n_2)$

$$P(n_i) = x \in \mathbf{R}^3, \quad |x| = 1, \quad < x, n_i >= 0, \quad i = 1, 2.$$

These planes divide sphere ω to four parts. One of these part (no matter which) we denote as Q. From the equality

$$\int_{\omega(n_1)} \mu^-(d\omega_p) - \int_{\omega(n_2)} \mu^-(d\omega_p) = 0$$

because of the oddness of μ^- $(\mu^-(-Q) = -\mu^-(Q))$ it follows that

$$\int_Q \mu^-(d\omega_p) = 0.$$

We restrict those Q, which lie in the half-space $x_3 \leq 0$. Let ω^- be a hemisphere which belongs to this half-space. Reflect it reciprocally unique to plane $x_3 = -1$ projecting from the origin. In this mapping region Q at sphere ω defines a rectilinear strip $T(Q)$ at plane $x_3 = -1$, and changing Q one can obtain any strip. The integral

$$\int_{T(Q)} J(x_1, x_2)\tilde{\mu}^-(dx_1, dx_2),$$

corresponds to the integral

$$\int_Q \mu^-(\mathrm{d}\omega_{\mathbf{p}}),$$

where J is the Jacobian of mapping by projecting. That is why from the equality

$$\int_Q \mu^-(\mathrm{d}\omega_{\mathbf{p}}) = 0$$

follows the equality

$$\int_{\alpha \leq <x,\mathbf{v}> \leq \beta} \tilde{\mu}^-(J\mathrm{d}x_1, \mathrm{d}x_2) = 0$$

for any vector $\mathbf{v} = (v_1, v_2)$ and for any α and β.

In this way, the single-valued solution of the equation brings about the uniqueness of the reconstruction of the measure by its meaning at half-spaces. A positive answer to this question is given in an article by (Khachaturov, 1954). The theorem is proved.

REMARK I. f $a \neq 0$, $b = 0$, or $a + b = 0$, then from equation (3.6) only the even part μ^+ of measure can be determined uniquely, and by $b \neq 0$, $a = 0$ the odd part.

Let in equation (3.6) measure μ be absolutely continuous, and $z(p)$ be density of measure μ. Consider the following equation with respect to $z(p)$:

$$f(n) = a \int_\omega |< n, p >| \, z(p)\mathrm{d}\omega_{\mathbf{p}} + b \int_\omega z(p)\mathrm{d}\omega_{\mathbf{p}} \equiv Az. \tag{3.10}$$

By Theorem 3.1 it follows, that equation (3.10) has no more than one solution $z(p) \in L_1(\omega)$.

Theorem 3.2. *The eigenvalues λ_r of equation (3.10) are*

$$\lambda_r = \begin{cases} 2\pi(a+b), & r = 0, \\ 4\pi a(-1)^k \dfrac{(|2k-3|)!!}{(|2k+2|)!!}, & r = 2k, \ k = 1, 2, \dots, \\ 2\pi b(-1)^k \dfrac{(|2k-1|)!!}{(|2k+2|)!!}, & r = 2k+1, \ k = 0, 1, \dots. \end{cases}$$

The eigenspace, corresponding to eigenvalue λ_r, consists of all spherical harmonics of order r.

PROOF. Take a set of spherical harmonics $Y_r(p)$ of order r in equation (3.10) and make sure that it satisfies the equality $AY = \lambda\lambda_r Y$, where

$$AY_r(n) = \int_\omega T(n,p) Y_r(p)\mathrm{d}\omega_{\mathbf{p}}.$$

Let $r = 0$, then $Y_0 = \text{const}$ and

$$AY_0 = a \int_\omega |< n, p >| \, Y_0 \mathrm{d}\omega_{\mathbf{p}} + b \int_{\omega(n)} Y_0 \mathrm{d}\omega_{\mathbf{p}} = 2\pi(a+b)Y_0.$$

If $r = 2k$, $k = 1, 2 \ldots$, then we have (see (Blaschke, 1916)):

$$\begin{aligned} a\int\limits_{\omega} |< n,p >| Y_{2k}(p) d\omega_{\mathbf{p}} + \frac{b}{2}\int\limits_{\omega} Y_{2k}(p) d\omega_{\mathbf{p}} &= a\int\limits_{\omega} |< n,p >| Y_{2k}(p) d\omega_{\mathbf{p}} \\ &= 4\pi a(-1)^k \frac{(|2k-3|)!!}{(2k+2)!!}. \end{aligned}$$

By $r = 2k+1$, $k = 0, 1 \ldots$ we obtain

$$AY_{2k+1} = a\int\limits_{\omega} |< n,p >| Y_{2k+1}(p) d\omega_{\mathbf{p}} + b\int\limits_{\omega(n)} Y_{2k+1}(p) d\omega_{\mathbf{p}}.$$

Since $Y_{2k+1}(p)$ is an odd function at ω, the first addend here is equal to zero. For calculating the second we accept point n as a pole of the spherical coordinate system γ, t, and

$$\begin{aligned} AY_{2k+1} &= b\int\limits_{\omega(n)} Y_{2k+1}(p) d\omega_{\mathbf{p}} \\ &= \sum_{m=0}^{2k+1} b\int\limits_{0}^{2\pi} d\tau \int\limits_{0}^{\frac{\pi}{2}} P^{m}_{2k+1}(\cos\gamma)\left[a^{m}_{2k+1}\cos m\tau + b^{m}_{2k+1}\sin m\tau\right]\sin\gamma d\gamma \\ &= 2\pi b a^{0}_{2k+1}\int\limits_{0}^{\frac{\pi}{2}} P_{2k+1}(\cos\gamma)\sin\gamma d\gamma \\ &= 2\pi b(-1)^k \frac{(|2k-1|)!!}{(2k+2)!!} a^{0}_{2k+1}, \end{aligned}$$

where $P^{(m)}_{k}$ are the associate Legendre polynomials, $P_k = P^{(0)}_{k}$. Here we seized on the equality of (Gradshtein and Ryzhik, 1980)

$$\int\limits_{0}^{1} P_{2k+1}(t) dt = (-1)^k \frac{(|2k-1|)!!}{(2k+2)!!}.$$

Show now that $a^{0}_{2k+1} = Y_{2k+1}$. For the average of the function Y_{2k+1} on the circumference $\gamma = \text{const}$ we have

$$\begin{aligned} \tilde{Y}_{2k+1}(\gamma, n) &= \frac{1}{2\pi}\int\limits_{0}^{2\pi} Y_{2k+1}(\gamma,\tau) d\tau \\ &= \frac{1}{2\pi}\sum_{m=0}^{2k+1}\int\limits_{0}^{2\pi} P^{(m)}_{2k+1}(\cos\gamma)\left[a^{m}_{2k+1}\cos m\tau + b^{m}_{2k+1}\sin m\tau\right] d\tau \\ &= P_{2k+1}(\cos\gamma) a^{0}_{2k+1}. \end{aligned}$$

Supposing in this equality $g = 0$ and considering that

$$\tilde{Y}_{2k+1}(\gamma, n)|_{g=0} = Y_{2k+1}, \quad P_{2k+1}(1) = 1,$$

we obtain $a^{0}_{2k+1} = Y_{2k+1}$.

In this way the spherical harmonics Y_r are eigenfunctions of operator $\mathbf{A}$. As the spherical harmonics form a basis in $L_2(w)$, the equation does not have any other eigenfunction. The theorem is proved.

Theorem 3.3. *If in equation (3.10) function $z(p)$ belongs to class $C^k(\omega)$, then function $f(n)$ belongs to class $C^{k+1}(\omega)$, $k = 0, 1 \dots$. Equation (3.10) is analytical on sphere ω solution $z(p)$ if and only if function $f(n)$ is analytical at ω.*

PROOF. Continue function $f(n)$ given on ω, to all values x, $x \neq 0$ supposing

$$f(x) = a \int_{\omega} |< x, p >| z(p) d\omega_p + b \int_{\langle x,p\rangle>0} z(p) d\omega_p, \quad x \in \mathbf{R}^3.$$

From

$$\frac{\partial f}{\partial x_i} = 2a \int_{\langle x,p\rangle>0} p_i z(p) d\omega_p + \frac{b}{|x|} \int_{\Gamma(x)} p_i z(p) ds_p, \quad i = 1, 2, 3,$$

where $\Gamma(x) = \{p \in \omega, < x, p >\}$ is the circumference at ω, it follows, that $f(x)$ belongs to class $C^1(\mathbf{R}^3 \setminus 0)$, if function $z(p)$ is continuous at ω. Let $z(p) \in C^1(\omega)$. Using (see (Blaschke, 1916; 1930))

$$\int_{\Gamma(x)} u(p) ds_p = \int_{\omega(n)} < \Delta_1(p_i z(p), p) - 2p_i z(p) p, x > d\omega_p \tag{3.11}$$

following from Green's formula it is possible to write

$$\int_{\Gamma(x)} p_i z(p) ds_p = \frac{1}{|x|} \int_{\langle x,p\rangle>0} < \Delta_1(p_i z(p), p) - 2p_i z(p) p, x > d\omega_p.$$

Here Δ_1 is the first differential Beltrami parameter. Then

$$\begin{aligned} \frac{\partial^2 f}{\partial x_i \partial x_j} &= \frac{1}{|x|} \int_{\Gamma(x)} p_i p_j z(p) ds_p - \frac{b x_j}{|x|^3} \int_{\Gamma(x)} p_i z(p) ds_p \\ &- \frac{b x_j}{|x|^4} \int_{\langle x,p\rangle>0} < \Delta_1(p_i z(p), p) - 2p_i z(p) p, x > d\omega_p \\ &+ \frac{b}{|x|^3} \int_{\Gamma(x)} < \Delta_1(p_i z(p)), p > p_j ds_p \\ &- \frac{b}{|x|^2} \int_{\langle x,p\rangle>0} < \Delta_1(p_i z(p), p) - 2p_i z(p) p, x > d\omega_p. \end{aligned}$$

From here it follows that $f(x)$ belongs to class $C^2(\mathbf{R}^3 \setminus 0)$. Consistently applying formula (3.11) to integral by $\Gamma(x)$ we obtain for $z(p) \in C^k(\omega)$ function $f(x) \in C^{k+1}(\mathbf{R}^3 \setminus 0)$ and, hence $f(n) \in C^{k+1}(\omega)$.

For the proof of the analyticity we use the following fact (see (Sobolev, 1974)): let on sphere ω a summable function $h(p)$ be given and let $h(p) = \sum_{k=0}^{\infty} Y_k(p)$ be its expansion in spherical harmonics. Then the function $h(p)$ is analytic on sphere ω if and only if

$$|Y_k(p)| < C \exp(-\nu k), \quad k = 0, 1, 2 \dots,$$

and $C > 0$, $\nu > 0$ are constants.

Suppose that $f(n)$ is an analytic function on ω and

$$f(n) = \sum_{k=0}^{\infty} Y_k(n), \qquad Y_k(n) = \frac{2k+1}{4\pi} \int_{\omega} f(p) P_k(< p, n >) d\omega_p,$$

is its expansion in spherical harmonics $Y_k(n)$.

From Theorem 3.2 it follows that it is possible to write the solution of equation (3.10) as a series:

$$z(n) = \frac{1}{2\pi(a+b)} Y_0 + \frac{1}{4\pi a} \sum_{k=0}^{\infty} (-1)^{k+1} \frac{(2k+2)!!}{(|2k-3|)!!} Y_{2k}(n)$$
$$+ \frac{1}{2\pi b} \sum_{k=0}^{\infty} (-1)^k \frac{(2k+2)!!}{(|2k-1|)!!} Y_{2k+1}(n)$$

Because of the fact mentioned above and $f(n)$ being analytic $|Y_k(p)| \le C \exp(-\nu k)$, $k = 0, 1, 2 \dots$, with certain positive constant C and ν. Therefore

$$\left| (-1)^{k+1} \frac{(2k+2)!!}{(|2k-3|)!!} Y_{2k}(n) \right| \le C_1 \mathrm{e}^{-2\nu_1 k},$$

$$\left| (-1)^k \frac{(2k+2)!!}{(|2k-1|)!!} Y_{2k+1}(n) \right| \le C_1 \mathrm{e}^{-(2k+1)\nu_1},$$

for a certain constant $C_1 > 0$ and $\nu_1 > 0$.

In this way, terms of expansion of the function $z(n)$ by spherical harmonics decrease exponentially. Hence $z(n)$ is analytic on ω. If $z(n)$ is an analytic function, $f(n)$ is analytic as well. The theorem is proved.

Consider an equation of the first kind with respect to the odd function $v(p)$, $p \in \omega$, on sphere ω:

$$h(n) = \frac{1}{2\pi} \int_{\omega(n)} v(p) d\omega_p. \tag{3.12}$$

Theorem 3.4. *If $h(n)$ is an odd function of class $C^3(\omega)$ at ω, then there exists a unique continuous solution of equation (3.12) and the formula of inversion holds:*

$$v(n) = -\frac{1}{2\pi} \int_{(n,p)>0} \frac{\Delta h(p)}{< n, p >} d\omega_p \tag{3.13}$$

PROOF. Suppose at first, that function $h(n)$ is an odd analytic function at ω. Expand it in a series in spherical harmonics:

$$h(n) = \sum_{k=0}^{\infty} Y_{2k+1}(n). \tag{3.14}$$

Since spherical harmonics Y_{2k+1} are eigenfunctions, and the corresponding eigenvalues are $(-1)^k \frac{(|2k-1|)!!}{(2k+2)!!}$ (see the proof of Theorem 3.2), the solution of equation (3.12) can be represented as series

$$v(p) = \sum_{k=0}^{\infty} (-1)^k \frac{(2k+2)!!}{(|2k-1|)!!} Y_{2k+1}(p), \tag{3.15}$$

and, as the terms of this serie decrease exponentially, it is an analytic function.

Set expansion (3.14) in integral

$$-\frac{1}{2\pi}\int\limits_{\langle n,p\rangle>0}\frac{\Delta h(p)}{< n,p >}\mathrm{d}\omega_{\mathbf{p}},$$

use equality

$$\Delta Y_{2k+1}(p)=-(2k+1)(2k+2)Y_{2k+1}(p),$$

integrate term by term, pass on to spherical coordinates γ, τ with respect to pole n and use equalities of (Gradshtein and Ryzhik, 1980):

$$Y_{2k+1}(\gamma,n)=P_{2k+1}(\cos\gamma)Y_{2k+1}(n),$$

$$\int\limits_0^1\frac{P_{2k+1}(t)}{t}\mathrm{d}t=(-1)^k\frac{(2k)!!}{(2k+1)!!},\quad k=0,1,2\ldots$$

We obtain

$$\begin{aligned}
-\frac{1}{2\pi}\int\limits_{\langle n,p\rangle>0}\frac{\Delta h(p)}{< n,p >}\mathrm{d}\omega_{\mathbf{p}} &= -\frac{1}{2\pi}\int\limits_{\langle n,p\rangle>0}\frac{\Delta\sum\limits_{k=0}^{\infty}Y_{2k+1}(p)}{< n,p >}\mathrm{d}\omega_{\mathbf{p}}\\
&= -\frac{1}{2\pi}\sum_{k=0}^{\infty}\int\limits_{\langle n,p\rangle>0}\frac{\Delta Y_{2k+1}(p)}{< n,p >}\mathrm{d}\omega_{\mathbf{p}}\\
&= -\frac{1}{2\pi}\sum_{k=0}^{\infty}(2k+1)(2k+2)\int\limits_{\langle n,p\rangle>0}\frac{Y_{2k+1}(p)}{< n,p >}\mathrm{d}\omega_{\mathbf{p}}\\
&= -\frac{1}{2\pi}\sum_{k=0}^{\infty}(2k+1)(2k+2)\int\limits_0^{\frac{\pi}{2}}\mathrm{d}\gamma\int\limits_0^{2\pi}\frac{Y_{2k+1}(\gamma,\tau)\sin\gamma}{\cos\gamma}\mathrm{d}\tau\\
&= \sum_{k=0}^{\infty}(2k+1)(2k+2)\int\limits_0^{\frac{\pi}{2}}\frac{P_{2k+1}(\cos\gamma)\sin\gamma}{\cos\gamma}\mathrm{d}\gamma Y_{2k+1}(n)\\
&= \sum_{k=0}^{\infty}(2k+1)(2k+2)(-1)^k\frac{(2k)!!}{(2k+1)!!}Y_{2k+1}(n)\\
&= \sum_{k=0}^{\infty}(-1)^k\frac{(2k+2)!!}{(|2k+1|)!!}Y_{2k+1}(n).
\end{aligned}$$

Compare the obtained series and expansion (3.15), and conclude that the solution of equation (3.12) is given by formula (3.13).

It is possible to represent solution $v(n)$ in the form

$$v(n)=h(n)-\int\limits_0^{\frac{\pi}{2}}\mathrm{tg}\ \gamma\tilde{h}''(\gamma,n)\mathrm{d}\gamma,\tag{3.16}$$

where $\tilde{h}''=\frac{\partial^2}{\partial\gamma^2}\tilde{h}(\gamma,n)$. To obtain (3.16) start with (3.13) in spherical coordinates γ, τ with respect to pole n. Using an expression for the Laplace–Beltrami operator in spherical coordinates, we have

$$\begin{aligned}
v(n) &= -\frac{1}{2\pi}\int\limits_{\langle n,p\rangle>0}\frac{\Delta h(p)}{<n,p>}\mathrm{d}\omega_{\mathbf{p}}\\
&= -\frac{1}{2\pi}\int\limits_0^{\frac{\pi}{2}}\mathrm{d}\gamma\int\limits_0^{2\pi}\frac{\Delta h(p(\gamma,\tau))}{\cos\gamma}\sin\gamma\mathrm{d}\tau\\
&= -\int\limits_0^{\frac{\pi}{2}}\mathrm{tg}\ \gamma\left[\frac{1}{2\pi}\int\limits_0^{2\pi}\Delta h(p(\gamma,\tau))\mathrm{d}\tau\right]\mathrm{d}\gamma\\
&= -\int\limits_0^{\frac{\pi}{2}}\mathrm{tg}\ \gamma\left\{\frac{1}{2\pi}\int\limits_0^{2\pi}\left[\frac{\partial^2 h}{\partial\gamma^2}+\mathrm{ctg}\ \gamma\frac{\partial h}{\partial\gamma}+\frac{1}{\sin^2\gamma}\frac{\partial^2 h}{\partial\tau^2}\right]\mathrm{d}\tau\right\}\mathrm{d}\gamma\\
&= -\int\limits_0^{\frac{\pi}{2}}\mathrm{tg}\ \gamma\tilde{h}''(\gamma,n)\mathrm{d}\gamma-\int\limits_0^{\frac{\pi}{2}}\mathrm{tg}\ \gamma\tilde{h}'(\gamma,n)\mathrm{d}\gamma-\frac{1}{2\pi}\int\limits_0^{\frac{\pi}{2}}\frac{\mathrm{tg}\ \gamma}{\sin^2\gamma}\left[\int\limits_0^{2\pi}\frac{\partial^2 h}{\partial\tau^2}\mathrm{d}\tau\right]\mathrm{d}\gamma\\
&= \tilde{h}(0,n)-\tilde{h}(\frac{\pi}{2},n)-\int\limits_0^{\frac{\pi}{2}}\mathrm{tg}\ \gamma\tilde{h}''(\gamma,n)\mathrm{d}\gamma\\
&\quad -\frac{1}{2\pi}\int\limits_0^{\frac{\pi}{2}}\frac{\mathrm{tg}\ \gamma}{\sin^2\gamma}\left[\frac{\partial h(\gamma,2\pi)}{\partial\tau}-\frac{\partial h(\gamma,0)}{\partial\tau}\right]\mathrm{d}\gamma.
\end{aligned}$$

Since

$$\frac{\partial h(\gamma,2\pi)}{\partial\tau}-\frac{\partial h(\gamma,0)}{\partial\tau}=0,\quad \tilde{h}(0,n)=h(n),$$

and because of the oddness of function $h(n)$ average $\tilde{h}\left(\frac{\pi}{2},0\right)=0$, finally we obtain

$$v(n)=h(n)-\int\limits_0^{\frac{\pi}{2}}\mathrm{tg}\ \gamma\tilde{h}''(\gamma,n)\mathrm{d}\gamma.$$

If function $h(n)$ belongs to class $C^3(\omega)$, then approximating it in norm of space $C^3(\omega)$ by a sequence of analytic functions, we establish the correctness of the theorem.

Theorem 3.5. *If in equation (3.10)* $a\neq 0$, $b\neq 0$, $a+b\neq 0$ *and function* $f(n)$ *belongs to class* $C^k(\omega)$, $k>4$, *then the formula of inversion holds*

$$\begin{aligned}
z(n) &= -\frac{1}{8\pi^2}\frac{\partial}{\partial t}\int\limits_{(n,p)^2>0}\frac{(\Delta+2)f_1(p)\,|<n,p>|}{\sqrt{<n,p>^2-t}}\mathrm{d}\omega_{\mathbf{p}}\Bigg|_{t=0}\\
&\quad -\frac{1}{4\pi^2 b}\int\limits_{\omega(n)}\frac{\Delta f^-(p)}{<n,p>}\mathrm{d}\omega_{\mathbf{p}}.
\end{aligned}\tag{3.17}$$

PROOF. As in the proof for the even and odd parts of the function $z(p)$ in Theorem 3.1, we obtain equations

$$f_1(p)=\int\limits_{\omega}|<n,p>|\,z^+(p)\mathrm{d}\omega_{\mathbf{p}},\tag{3.18}$$

$$f^-(n) = b \int\limits_{\omega(n)} z^{-1}(p)\mathrm{d}\omega_{\mathbf{p}}. \tag{3.19}$$

It is known that equation (3.18) is uniquely solvable and its solution is given by (Blaschke, 1916; Pogorelov, 1976)

$$\begin{aligned} z^+(n) &= -\frac{1}{8\pi^2}\frac{\partial}{\partial t} \int\limits_{(n,p)^2>t} \frac{(\Delta+2)f_1(p)\,|<n,p>|}{\sqrt{<n,p>^2-t}}\mathrm{d}\omega_{\mathbf{p}}\Bigg|_{t=0} \\ &= \frac{1}{4\pi}(\Delta+2)f_1(p) + \frac{1}{4\pi}\int\limits_0^{\frac{\pi}{2}} \frac{\left[(\Delta+2)\tilde{f}_1(\gamma,n)\right]'}{\cos\gamma}\mathrm{d}\gamma. \end{aligned} \tag{3.20}$$

In this way, the even part of the function is determined uniquely and constructively.

The odd part of $z^-(n)$ is determined uniquely from equation (3.19) by

$$z^-(n) = -\frac{1}{4\pi^2 b}\int\limits_{\omega(n)} \frac{\Delta f^-(p)}{<n,p>}\mathrm{d}\omega_{\mathbf{p}} = \frac{1}{2\pi b}f^-(n) - \frac{1}{2\pi b}\int\limits_0^{\frac{\pi}{2}} \mathrm{tg}\ \gamma \tilde{f}^{-\prime\prime}\mathrm{d}\gamma. \tag{3.21}$$

Summing (3.20) and (3.21) by equality $z(n) = z^+ + z^-$ we obtain (3.17).

REMARK I. f $a = 0$ or $a + b = 0$, then from equation (3.10) only the odd part $z^-(n)$ of the unknown function $z(n)$ is determined uniquely, and if $b = 0$, then only the even part $z^+(n)$ can be found uniquely.

Studying the inversion formula of equation (3.10) leads to the following result.

Theorem 3.6. *If $a \neq 0$, $b \neq 0$, $a+b \neq 0$ and function $f(n) \in C^k(\omega)$, $k > 4$, then solution $z(n)$ of equation (3.10) belongs to at least class $C^{k-4}(\omega)$ and the following estimate holds:*

$$\|z(n)\|_{C(\omega)} \leq M\,\|f(n)\|_{C^4(\omega)}, \tag{3.22}$$

where constant M depends only on a and b.

PROOF. By the previous theorem it is possible to write the solution of equation (3.10) in the form

$$\begin{aligned} z(n) = &\frac{1}{4\pi}(\Delta+2)f_1(p) + \frac{1}{4\pi}\int\limits_0^{\frac{\pi}{2}} \frac{\left[(\Delta+2)\tilde{f}_1(\gamma,n)\right]'}{\cos\gamma}\mathrm{d}\gamma \\ &+\frac{1}{2\pi b}f^-(n) - \frac{1}{2\pi b}\int\limits_0^{\frac{\pi}{2}} \mathrm{tg}\ \gamma\tilde{f}^{-\prime\prime}\mathrm{d}\gamma. \end{aligned}$$

Since function $f_1(n)$ is even on ω, equality $\left[(\Delta+2)\tilde{f}_1(\gamma,n)\right]'\Big|_{\gamma=\frac{\pi}{2}} = 0$ is fulfilled. Using this equality and the existence of a continuous derivative $\left[(\Delta+2)\tilde{f}_1(\gamma,n)\right]''$ we can represent it as follows:

$$\left[(\Delta+2)\tilde{f}_1(\gamma,n)\right]' = \left(\gamma-\frac{\pi}{2}\right)\int\limits_0^1 \left[(\Delta+2)\tilde{f}_1\left(t\left(\gamma-\frac{\pi}{2}\right)+\frac{\pi}{2},n\right)\right]''\mathrm{d}t.$$

In the same way, because of the oddness of function $f^-(n)$ to ω, $\tilde{f}^{-\prime\prime}(\gamma,n)\Big|_{\gamma=\frac{\pi}{2}} = 0$ and

$$\tilde{f}^{-\prime\prime}(\gamma,n) = \left(\gamma - \frac{\pi}{2}\right)\int\limits_0^1 \tilde{f}^{-\prime\prime\prime}\left(t\left(\gamma-\frac{\pi}{2}\right)+\frac{\pi}{2},n\right)\mathrm{d}t$$

hold. Consequently

$$z(n) = \frac{1}{4\pi}(\Delta+2)f_1(n) + \frac{1}{4\pi}\int\limits_0^{\frac{\pi}{2}}\frac{\gamma-\frac{\pi}{2}}{\cos\gamma}\left\{\int\limits_0^1\left[(\Delta+2)\tilde{f}_1\left(t\left(\gamma-\frac{\pi}{2}\right)+\frac{\pi}{2},n\right)\right]''\mathrm{d}t\right\}\mathrm{d}\gamma$$

$$+\frac{1}{2\pi b}f^-(n) + \frac{1}{2\pi b}\int\limits_0^{\frac{\pi}{2}}\left(\frac{\pi}{2}-\gamma\right)\operatorname{tg}\gamma\left[\tilde{f}^{-\prime\prime\prime}\left(t\left(\gamma-\frac{\pi}{2}\right)+\frac{\pi}{2},n\right)\mathrm{d}t\right]\mathrm{d}\gamma. \tag{3.23}$$

From the last formula it follows that $z(n)$ belongs to class $C^{k-4}(\omega)$.

Prove the estimate. Using again formula (3.23), we have

$$\begin{aligned}\|z(n)\|_{C(\omega)} &\le \frac{1}{4\pi}\|(\Delta+2)f_1(n)\|_{C(\omega)} + \frac{1}{4\pi}\int\limits_0^{\frac{\pi}{2}}\frac{\frac{\pi}{2}-\gamma}{\cos\gamma}\mathrm{d}\gamma \\ &\times \sup_{\substack{0\le\gamma\le\frac{\pi}{2}\\ h\in\omega}}\left|\left[(\Delta+2)\tilde{f}_1\right]''\right| + \frac{1}{2\pi|b|}\left\|f^-(n)\right\|_{C(\omega)} \\ &+\frac{1}{2\pi|b|}\int\limits_0^{\frac{\pi}{2}}\left(\frac{\pi}{2}-\gamma\right)\operatorname{tg}\gamma\mathrm{d}\gamma\sup_{\substack{0\le\gamma\le\frac{\pi}{2}\\ h\in\omega}}\left|\tilde{f}^{-\prime\prime\prime}\right|.\end{aligned}$$

Estimate every addend in the right-hand part of this inequality:

$$\|(\Delta+2)f_1(n)\|_{C(\omega)} \le \frac{1}{|a|}\|\Delta f\|_{C(\omega)} + \frac{2}{|a|}\|f\|_{C(\omega)} + \frac{2|b|}{|a||a+b|}\|f\|_{C(\omega)},$$

$$\sup_{\gamma,n}\left|\left[(\Delta+2)\tilde{f}_1\right]''\right| \le \frac{1}{|a|}\sup_{\gamma,n}\left|\Delta\tilde{f}''\right| + \frac{2}{|a|}\sup_{\gamma,n}\left|\tilde{f}''\right|,$$

$$\sup_{\gamma,n}\left|\tilde{f}^{-\prime\prime\prime}\right| \le \sup_{\gamma,n}\left|\tilde{f}'''\right|,$$

$$\frac{1}{2\pi b}\int\limits_0^{\frac{\pi}{2}}\left(\frac{\pi}{2}-\gamma\right)\operatorname{tg}\gamma\mathrm{d}\gamma = \frac{\pi}{2}\ln 2, \qquad \int\limits_0^{\frac{\pi}{2}}\frac{\frac{\pi}{2}-\gamma}{\cos\gamma}\mathrm{d}\gamma = 2G.$$

G is the Catalan constant. From these reduced inequalities the estimate follows:

$$\begin{aligned}\|zn\|_{C(\omega)} &\le \frac{1}{2\pi}\left(\frac{1}{|a|}+\frac{1}{|b|}+\frac{|b|}{|a||a+b|}\right)\|f(n)\|_{C(\omega)} + \frac{1}{4\pi|a|}\|\Delta f\|_{C(\omega)} \\ &+\frac{G}{2\pi|a|}\sup_{\gamma,n}\left|\Delta\tilde{f}''\right| + \frac{G}{\pi|a|}\sup_{\gamma,n}\left|\tilde{f}''\right| + \frac{\ln 2}{4\pi|a|}\sup_{\gamma,n}\left|\tilde{f}'''\right| \\ &\le \frac{1}{2\pi|a|}\left(1+\frac{|a|}{|b|}+\frac{b}{|a+b|}\right)\|f\|_{C(\omega)} + \frac{1}{4\pi|a|}\|f\|_{C^2(\omega)} \\ &+\frac{G}{2\pi|a|}\|f\|_{C^4(\omega)} + \frac{G}{\pi|a|}\|f\|_{C^2(\omega)} + \frac{\ln 2}{4\pi|b|}\|f\|_{C^3(\omega)} \\ &\le M(a,b)\|f\|_{C^4(\omega)}.\end{aligned}$$

The theorem is proved.

On the basis of the inversion formula of equation (3.10) by a standard manner more general integral equations of the first kind may be examined, including nonlinear ones. Consider, for instance,

$$f(n) = \int\limits_{\omega} \mathbf{P}(n,p)z(p)\mathrm{d}\omega_{\mathbf{p}} + \mathbf{T}z,$$

where $\mathbf{P}(n,p) = a\,|<n,p>| + b\chi(<n,p>)$, and $\mathbf{T}$ be a certain operator, using the inversion formula, we have

$$z(n) = \bar{f}(n) + \mathbf{B}z, \tag{3.24}$$

where $\bar{f} = \mathbf{P}^{-1}f$, $\mathbf{B} = \mathbf{P}^{-1}\mathbf{T}$. Relation (3.24) is an equation of the second kind and, for instance, by corresponding restriction to $\mathbf{T}$ operator $\mathbf{B}$ may be contracting, which allows us to use methods of the theory of fixed points.

Another application of the inversion formula can be connected to the problem of integral geometry. Consider, for example, the following problem: to find function $u(n) \in C^{\infty}(\omega)$, $n \in \omega$, if function $\varphi(n)$ is known:

$$\varphi(n) = 2a \int\limits_{\Gamma(n)} u(p)\mathrm{d}s + b \int\limits_{\omega(n)} (\Delta + 2)u(p)\mathrm{d}\omega_{\mathbf{p}}. \tag{3.25}$$

In the case $a \neq 0$, $b = 0$, we obtain the problem about the recovery of $u(p)$ by its integral along large a circle sphere. Minkowski showed through the use of expansion in spherical harmonics, that the even part of $u(p)$ is uniquely determined for $b = 0$. Later Funk obtained a solution of this problem in a closed form, using the inversion formula of the Abel integral equation (see (Funk, 1916)).

Theorem 3.7. *Let $a \neq 0$, $b \neq 0$, $a + b \neq 0$, $v(n) \in C^{\infty}(\omega)$ and $\int\limits_{\omega} n\varphi(n)\mathrm{d}\omega = 0$. Then there exists a solution $u(p)$ of equation (3.25), $u(p) \in C^{\infty}(\omega)$ and inversion formula holds*

$$u(p) = \frac{1}{4\pi} \int\limits_{\omega} \Phi(p)\left\{(1- <n,p>)\ln(1- <n,p>)\right\}\mathrm{d}\omega_{\mathbf{p}},$$

where

$$\begin{aligned} \Phi(n) \;&=\; -\frac{1}{8\pi^2}\frac{\mathrm{d}}{\mathrm{d}t} \int\limits_{(n,p)^2>t} \frac{(\Delta+2)\varphi_1(p)\,|<n,p>|}{\sqrt{<n,p>^2 - t}}\mathrm{d}\omega_{\mathbf{p}}\Bigg|_{t=0} \\ &\quad - \frac{1}{4\pi b}\int\limits_{\omega(n)} \frac{\Delta\varphi^{-}(p)}{<n,p>}\mathrm{d}\omega_{\mathbf{p}}. \end{aligned}$$

Proof. . With the use of Green's formula we obtain

$$2\int\limits_{\Gamma(n)} u(p)\mathrm{d}s = \int\limits_{\omega(n)} |<n,p>|\,(\Delta+2)u(p)\mathrm{d}\omega_{\mathbf{p}}. \tag{3.26}$$

Therefore equation (3.25) with respect to function $u(p)$ in accordance to (3.25) and (3.26) acquires the form

$$\varphi(n) = \int\limits_{\omega} \left\{a\,|<n,p>| + b\chi(<n,p>)\right\}(\Delta+2)u(p)\mathrm{d}\omega_{\mathbf{p}}. \tag{3.27}$$

The solution of equation (3.27) exists and is unique, if

$$\int_{\omega} n\varphi(n)\mathrm{d}\omega_n = 0. \tag{3.28}$$

Indeed, let

$$\varphi(n) = \sum_{k=0}^{\infty} Y_k(n), \quad Y_k(n) = \frac{2k+1}{4\pi}\int_{\omega} \varphi(p)P_k(< n,p >)\mathrm{d}\omega_{\mathrm{p}}$$

be a series of expansion in spherical harmonics of the left-hand part of equation (3.27). Because of (3.28) for the first harmonic we find $Y_1(n) = 0$. Then function

$$u(p) = \sum_{\substack{r=0\\ r\neq 1}}^{\infty} \frac{\lambda_r^{-1}}{(r-1)(r+2)} Y_r(p),$$

where eigenvalues λ_r of equation (3.10) are solutions of equation (3.27). If $\varphi(n)$ is infinitely differentiable on ω, then function $\Delta^l\varphi$ is bounded to ω, for all values of l of operator Δ. That is why, and from the estimate of the derivative of spherical harmonics (Sobolev, 1974)

$$|\Delta^\alpha Y_r(p)| \leq C r^{m+1-2l} \left\|\Delta^l\varphi\right\|_{L_2(\omega)}, \quad |\alpha| = m,$$

$C > 0$ is constant, infinite differentiability of solution $u(p)$ follows.

Consider equation (3.27) as an integral equation of the first kind with respect to $(\Delta+2)u(p)$. According to the inversion formula of equation (3.10) we find

$$\begin{aligned}(\Delta+2)u(n) = \Phi(n) = & -\frac{1}{8\pi^2}\frac{\mathrm{d}}{\mathrm{d}t}\int_{(n,p)^2>t} \frac{(\Delta+2)\varphi_1(p)\,|< n,p >|}{\sqrt{< n,p >^2 - t}}\mathrm{d}\omega_{\mathrm{p}}\Bigg|_{t=0} \\ & -\frac{1}{4\pi^2 b}\int_{\omega(n)} \frac{\Delta\varphi^-(p)}{< n,p >}\mathrm{d}\omega_{\mathrm{p}}.\end{aligned} \tag{3.29}$$

Applying the Weingarten formula (Blaschke, 1930; Pogorelov, 1973), from (3.29), we obtain the wanted result:

$$u(p) = \frac{1}{4\pi}\int_{\omega} \Phi(p)\{(1- < n,p >)\ln(1- < n,p >)\}\mathrm{d}\omega_{\mathrm{p}}.$$

3.3 UNIQUENESS

Here some results are given on the univalent definiteness of a closed convex surface, if the functionals of its orthogonal projections and illuminated parts are known. The investigation method consists of the following: at first the problem has to be reduced to the integral equation (3.1), and then the results are used to solvs the problem of Minkowski and Christoffel. All surfaces are, as a rule, supposed to be strictly convex and sufficiently smooth.

Consider the function

$$\varphi(n) = 2aF(n) + bS(n),$$

where $F(n)$ is the area of orthogonal projection of the closed convex surface B on the plane $P(n) = \{x \in \mathbf{R}^3, < x, n >= 0\}$, $S(n)$ is the area of the illuminated part of the surface in the direction n, a and b are constants.

$F(n)$ and $S(n)$, can be represented as

$$F(n) = \frac{1}{2}\int\limits_{\omega} |< n, p >| G(B, d\omega_p), \quad S(n) = \int\limits_{\omega(n)} G(B, d\omega_p),$$

where $G(B,\cdot)$ is a surface function. Using these representations it is possible write function $\varphi(n)$ as

$$\varphi(n) = a\int\limits_{\omega} |< n, p >| G(B, d\omega_p) + b\int\limits_{\omega(n)} G(B, d\omega_p). \tag{3.30}$$

A closed convex surface is defined uniquely, apart from a translation, by setting the surface function (Aleksandrov, 1937a,b; 1938a,b). Therefore and because of the uniqueness of the solution of equation (3.30) (see Theorem 3.1) the following theorem takes place.

Theorem 3.8. *A closed convex surface is defined uniquely except for a translation by function $\varphi(n)$, $n \in \omega$, $a \neq 0$, $b \neq 0$, $a + b \neq 0$.*

Let $G_1(B,\cdot)$ be the first function of curvature of a closed convex surface B, let $L(n)$ be the length of a boundary orthogonal projection of the surface B on the plane $P(n)$, $M(n)$ is the average integral curvature of the illuminated part of the surface in the direction n. Functions $L(n)$ and $M(n)$ are determined by

$$L(n) = \frac{1}{2}\int\limits_{\omega} |< n, p >| G_1(B, d\omega_p), \quad M(n) = \frac{1}{2}\int\limits_{\omega(n)} G_1(B, d\omega_p).$$

Consider the function

$$\psi(n) = 2aL(n) + 2bM(n),$$

where a and b are constants, $a \neq 0$, $b \neq 0$, $a+b \neq 0$. Through the use of the representation for $L(n)$ and $M(n)$ (3.30) can be written in the form

$$\psi(n) = a\int\limits_{\omega} |< n, p >| G_1(B, d\omega_p) + b\int\limits_{\omega(n)} G_1(B, d\omega_p). \tag{3.31}$$

Expression (3.31) is an equation of the first kind with respect to the function of curvature $G_1(B,\cdot)$. The solution of the given equation is unique using Theorem 3.1, but since a closed convex surface is defined uniquely, except for a translation, by setting first the function curvature, the following theorem is equitable.

Theorem 3.9. *A closed convex surface B is defined uniquely apart from a translation by function $\psi(n)$, $n \in \omega$.*

REMARK I. f surface B is centrally-symmetric, then Theorems 3.8 and 3.9 are equitable at $b = 0$, that is, a convex centrally-symmetric surface B is uniquely defined by the area and length of its orthogonal projection on a plane in all directions. These are well-known facts (see (Blaschke, 1916)). A noncentrally symmetric convex surface is not uniquely defined by the area of its orthogonal projection. So, a piecewise analytic convex surface of revolution, which form has parametric representation

$$x = \frac{A}{\sqrt{2}}\cos\varphi, \quad y = 0, \qquad x = A\cos\left(\frac{\pi}{4} - \varphi\right), \quad y = 0,$$

$$z = A\int_0^{\varphi}\sqrt{1 - \frac{\sin^2\varphi}{2}}\mathrm{d}\varphi, \quad 0 \le \varphi \le \frac{\pi}{2},$$

$$z = A\left[\frac{1}{\sqrt{2}} - \sin\left(\frac{\pi}{4} - \varphi\right)\right], \quad -\frac{\pi}{4} \le \varphi \le 0,$$

is, when being a sphere, a surface of constant luminance, that is, it has a constant area of projection on any plane (see (Blaschke, 1916)). A convex noncentrally-symmetric surface is also not defined uniquely by the length boundaries of its orthogonal projection. For example, a convex surface, having support function $H(x) = h(x) + |x|$, where $h(x)$ is an odd function, is a surface of constant coverage, i.e., it has a constant length of projection on any plane.

Theorems 3.8 and 3.9 establish the uniqueness of reconstruction of the convex surface B, but do not give a constructive method to determine its form. When surface B is sufficiently smooth, the problem of determining its form by $\varphi(n)$ or $\psi(n)$ is reduced to the inversion of equations (3.30), (3.31) and to solve the equation in partial derivatives of the second order for the support function of the surface.

Let $R_1(n)$, $R_2(n)$ be the principal radii of curvature of a closed strictly convex, and at least twice continuously differentiable surface B as function of unit vector n normal to B. In such cases the surface function $G(B,\cdot)$ has a continuous density $z(n) = R_1(n)R_2(n)$ and $\varphi(n)$ is written in as

$$v(n) = a\int_{\omega} |\langle n,p\rangle|\, R_1(p)R_2(p)\mathrm{d}\omega_p + b\int_{\omega(n)} R_1(p)R_2(p)\mathrm{d}\omega_p. \tag{3.32}$$

Consider (3.32) as an equation of the first kind with respect to the product of the principal radii of curvature $R_1(p)R_2(p)$. Using Theorem 3.5 leads to following result:

Theorem 3.10. *Let $a \neq 0$, $b \neq 0$, $a + b \neq 0$, let B be a closed convex regular surface. Then:*

$$\begin{aligned} R_1(n)R_2(n) = & -\frac{1}{8\pi^2}\frac{\mathrm{d}}{\mathrm{d}t}\int_{\langle n,p\rangle^2 > t} \frac{(\Delta+2)\varphi_1(p)\,|\langle n,p\rangle|}{\sqrt{\langle n,p\rangle^2 - t}}\mathrm{d}\omega_p\Bigg|_{t=0} \\ & -\frac{1}{4\pi^2 b}\int_{\omega(n)}\frac{\Delta\varphi^-(p)}{\langle n,p\rangle}\mathrm{d}\omega_p, \end{aligned}$$

$$R_1(n)R_2(n) = \begin{vmatrix} H_{11}(n) & H_{12}(n) \\ H_{21}(n) & H_{22}(n) \end{vmatrix} + \begin{vmatrix} H_{22}(n) & H_{23}(n) \\ H_{32}(n) & H_{33}(n) \end{vmatrix} + \begin{vmatrix} H_{33}(n) & H_{13}(n) \\ H_{31}(n) & H_{11}(n) \end{vmatrix}. \tag{3.33}$$

Here $H_{ij}(n)$ are the derivatives $\dfrac{\partial^2 H(x)}{\partial x_i \partial x_j}$ of the support function $H(x)$, calculated at $x = n \in \omega$.

From the theorem it follows, that the recovering of a regular surface B by the function $\varphi(n)$ is reduced to the solution of equation (3.33) for its support function. Formula (3.33) is well-known (Bakel'man et al., 1973).

At the conditions to surface B as above, the first function of curvature $G_1(B,\cdot)$ has a density, which is equal to the sum of the principal radii of curvature. That is why it is possible to write function $\psi(n)$ in the form

$$\psi(n) = a\int_{\omega} |< n,p >| \, [R_1(p) + R_2(p)] \mathrm{d}\omega_{\mathrm{p}} + b \int_{\omega(n)} [R_1(p) + R_2(p)] \mathrm{d}\omega_{\mathrm{p}}.$$

Theorem 3.11. *Let $a \neq 0$, $b \neq 0$, $a + b \neq 0$, and B be a closed strictly convex regular surface. Then:*

$$\begin{aligned} R_1(n)R_2(n) = & - \frac{1}{8\pi^2}\frac{\mathrm{d}}{\mathrm{d}t} \int_{<n,p>^2>t} \frac{(\Delta + 2)\psi_1(p) \, |< n,p >|}{\sqrt{< n,p >^2 - t}} \mathrm{d}\omega_{\mathrm{p}} \Bigg|_{t=0} \\ & - \frac{1}{4\pi^2 b} \int_{\omega(n)} \frac{\Delta\psi^-(p)}{< n,p >} \mathrm{d}\omega_{\mathrm{p}}, \end{aligned} \tag{3.34}$$

$$H(n) = \frac{1}{4\pi} \int_{\omega} [R_1(p) + R_2(p)] \left\{(1- < n,p >) \ln(1- < n,p >)\right\}.$$

Using (3.34) the question recovering regular surface B by the function $\psi(n)$ is solved constructively in explicit form (Materon, 1975).

Consider now the problem of determining a closed convex surface and a function, given at its surface, on the condition that the functionals from its orthogonal projection and illuminated portion are given. Similar problems appear also in inverse problems of scattering.

Let B be a closed strictly convex smooth surface in $\mathbf{R}^3$ and let $u(n)$ be a continuous function, given at B as a function of the unit normal vector of this surface. Consider functions

$$\varphi_1(p) = a_1 \int_{\omega} |< n,p >| \, u(p)R_1(p)R_2(p)\mathrm{d}\omega_{\mathrm{p}} + b_1 \int_{\omega(n)} u(p)R_1(p)R_2(p)\mathrm{d}\omega_{\mathrm{p}}, \tag{3.35}$$

$$\varphi_2(p) = a_2 \int_{\omega} |< n,p >| \, u(p) \left[R_1(p) + R_2(p)\right] \mathrm{d}\omega_{\mathrm{p}} + b_2 \int_{\omega(n)} u(p) \left[R_1(p) + R_2(p)\right] \mathrm{d}\omega_{\mathrm{p}}, \tag{3.36}$$

where R_1, R_2 are the principal radii of curvature, a_i, b_i are constants, $a_i \neq 0$, $b_i \neq 0$, $a_i + b_i \neq 0$, $i = 1, 2$. The geometrical sense of the introduced function at $u(p) = 1$ are described above.

Theorem 3.12. *A closed strictly convex, twice continuously differentiable surface B and a continuous function $u(n)$, $u(n) > 0$, $n \in \omega$, are uniquely defined by functions $v_1(n)$, $v_2(n)$, $n \in \omega$ (surface B can be restored up to a translation).*

PROOF. Because of the uniqueness of the solution of equations (3.35) and (3.36) functions $\Phi_1(n) = u(n)R_1R_2$, $\Phi_2(n) = u(n)[R_1 + R_2]$ are uniquely found by functions $\varphi_1(n)$ and $\varphi_2(n)$, respectively. Considing that $u(n) > 0$ at ω we obtain equality

$$\frac{R_1(n)R_2(n)}{R_1(n) + R_2(n)} = \Phi(n),$$

where $\Phi(n) = \Phi_1/\Phi_2$ is a given function. It is known that this function defines surface B uniquely up to a translation (Pogorelov, 1973). That is why the principal radii of curvature R_1 and R_2 are uniquely established. With equality $\Phi_1 = uR_1R_2$ we find function $u(n)$. The theorem is proved.

In scattering or emission problem one has to determine both the scatterer $B(t)$, the form of which can vary in the course of time t, as well as the function, characterizing the density of the source on it. Taking this into account consider equation

$$\Phi(n,t) = \int\limits_{B(t)} P(n,p)u(p)\mathrm{d}s_t \tag{3.37}$$

where $P(n,p) = a\,|< n,p >| + b\chi(< n,p >)$, $a \neq 0$, $b \neq 0$, $a + b \neq 0$, $\mathrm{d}s_t$ is an element of area $B(t)$.

Theorem 3.13. *Let surface $B(t)$ be closed at any $t > 0$, strictly convex, twice continuously differentiable, and let the following equality hold:*

$$\frac{\mathrm{d}}{\mathrm{d}t}\left[R_1(p,t)R_2(p,t)\right]\Big|_{t=0} = \alpha\left[R_1(p,0) + R_2(p,0)\right],$$

where $R_1(p,t)$, $R_2(p,t)$ are the principal radii of curvature of surface $B(t)$, $t > 0$, $a \neq 0$ is a constant. Then the continuous function $u(p) > 0$ and the form of surface $B(t)$ for any $t \geq 0$ are uniquely defined by function $\Phi(n,t)$, $n \in \omega$, $t > 0$.

PROOF. Consider relation $\mathrm{d}s_t = R_1(p,t)R_2(p,t)\mathrm{d}\omega_p$ equitable for smooth surface, and rewrite equation (3.37) as

$$\Phi(n,t) = \int\limits_{\omega} P(n,p)u(p)R_1(p,t)R_2(p,t)\mathrm{d}\omega_p. \tag{3.38}$$

Differentiating equation (3.38) with respect to t and using the conditions of the theorem, we obtain

$$\frac{\mathrm{d}\Phi}{\mathrm{d}t}\Big|_{t=0} = a\int\limits_{\omega} P(n,p)u(p)\left[R_1(p,0) + R_2(p,0)\right]\mathrm{d}\omega_p. \tag{3.39}$$

Function $W_1(n,t) = u(n)R_1(n,t)R_2(n,t)$ from equation (3.38) is uniquely defined by function $\Phi(n,t)$, but from equation (3.39) function $W_2(n) = u(n)\left[R_1(n,0) + R_2(n,0)\right]$ is uniquely defined by function $\Phi_t'(n,0)$. From this surface $B(0)$ is uniquely defined exept for a translation by the function

$$W(n) = \frac{R_1(n,0)R_2(n,0)}{R_1(n,0) + R_2(n,0)}.$$

Then the principal radii of curvature $R_1(n,0)$ and $R_2(n,0)$ of surface $B(0)$ are uniquely found. That is why from $W_2(n) = u(n)\left[R_1(n,0) + R_2(n,0)\right]$ we find $u(n)$. Knowing $u(n)$

from $W_1(n,t) = u(n)R_1(n,t)R_2(n,t)$ we define $R_1(n,t)R_2(n,t)$. Using the uniqueness of the solution of the Minkowski problem by function W_1/u we uniquely define surface $B(t)$ for any $t > 0$, apart from a translation. The theorem is proved.

Let the area and length of a boundary of an orthogonal projection of convex surface B on plain in all directions be given. The possibility to define surface B by the functions $F(n)$, $n \in \omega$, $L(n)$, $n \in \omega$ is given by the following theorem.

Theorem 3.14. *Let B_1 and B_2 be closed convex analytic surfaces in $\mathbf{R}^3$. If $F_1(n) = F_2(n)$, $L_1(n) = L_2(n)$ for any $n \in \omega$ then surfaces B_1 and B_2 are equal.*

PROOF. Area $F(n)$ and length $L(n)$ of the boundary of a closed convex, at least twice continuously differentiable surface B are defined by

$$F(n) = \frac{1}{2}\int_\omega [R_1(p)R_2(p)]^+ \,|< n,p >|\,\mathrm{d}\omega_{\mathbf{p}}, \tag{3.40}$$

$$L(n) = \frac{1}{2}\int_\omega [R_1(p)+R_2(p)]^+ \,|< n,p >|\,\mathrm{d}\omega_{\mathbf{p}}, \tag{3.41}$$

where R_1 and R_2 are the principal radii of curvature as functions of the unit normal vector. Let $H(x)$ be a support function of the surface B. It is well known that R_1 and R_2 are eigenvalues of matrix $\|\mathbf{H}_{ij}\|$ of the second partial derivatives of the function $H(x)$. That is why we can write

$$\Delta H = -(R_1 + R_2),$$

from which we obtain that

$$\Delta H^+ = -(R_1 + R_2)^+. \tag{3.42}$$

here Δ is the Laplace operator. Hence, if function $L(n)$ is given, then from equation (3.41) $(R_1 + R_2)^+$ is uniquely defined, and from the equation above the even part of support function $H(x)$ is uniquely defined (Christoffel theorem).

In order to prove the theorem, it is sufficient to establish that from setting $F(n)$ (or from setting $[R_1(n)R_2(n)]^+$) the unique definiteness of H^- follows up to an arbitrary linear function (addition of a linear function to H means a new choice of the origin of the coordinate system).

Since R_1R_2 is equal to the sum of main minors of matrix $\|\mathbf{H}_{ij}\|$, then filling $\mathbf{H}_{ij}$ as a sum of even and odd parts and picking out parameters u and v on sphere ω by specific manner (Buseman, 1958), for odd part, H^-, of the support function we obtain equation

$$H^-_{uu}H^-_{vv} - {H^-_{uv}}^2 = \psi(u,v), \tag{3.43}$$

where $\psi(u,v)$ is known function.

Our problem is the definition of H^- up to a linear function and multiplication by (-1) (this is equivalent to rolling of surface B) from equation (3.43).

Show, that there exists a point (u_0, v_0) such that $\psi(u_0,v_0) > 0$ (it is assumed that surface B does not have a center of symmetry and therefore H^- is non-linear).

Since H^- is an odd function, there exists a closed curve $l = (u(t), v(t))$ such that in domain D, the bounded curve l is simply connected and $H^- = 0$ on l. Therefore, if everywhere on ω function $\psi(u,v) \leq 0$, then in a domain D we would obtain surface $H^- = H^-(u,v)$ of nonpositive curvature, equal zero at l, which is impossible.

Let (u_0, v_0) be a point, in which $\psi > 0$. Let K denote a circle of certain radius centered in point (u_0, v_0), in which $\psi > 0$. Because of the continuity of $\psi(u,v)$ such a circle exists.

If now H_1 and H_2 are support functions of surfaces B_1 and B_2 and convex in the circle in one side (it is always possible to roll one of the surfaces), then adding, for instance, a certain linear function to H_1, one can obtain $H_1 = H_2$ on a certain closed curve l_1, lying inside circle K. Since $H_1^+ = H_2^+$ everywhere on ω, then $H_1^- = H_2^-$ on l_1.

That is why, taking into account that H_1^- and H_2^- satisfy the same equation (3.43) and furthermore $H_1^- = H_2^-$ at l_1, then $H_1^- = H_2^-$ everywhere in the domain, bounded by l_1 (Bakel'man, 1965). And by supposing B_1 and B_2 analytic everywhere on the sphere $H_1^- = H_2^-$. The theorem is proved.

REMARK. (Campi, 1986) giives un example showing that the condition of analyticity in this theorem is essential.

In 1926 (Matzumara, 1926) proved, that a closed convex surface, of which area and length of boundaries of its orthogonal projections on any plane are constant, is a sphere.

In connection with this result it is naturally to expect that if isoperimetric defect $L^2 - 4\pi F(n)$ is small for each isoperimetric projection $B(n)$ of surface B, then surface B differs a little from a sphere. Let R and r be radii of described and inscribed balls of convex surface.

Theorem 3.15. *If $L^2 - 4\pi F(n) \leq \varepsilon^2$, $n \in \omega$, then $R - r < \dfrac{\varepsilon}{\pi}$.*

PROOF. We again use representations

$$F(n) = -\frac{1}{2}\int\limits_{\omega} |<n,p>| G(B, d\omega_p), \quad L(n) = \int\limits_{\Gamma(n)} H(p) ds_p$$

for the area and the length of the boundary of an orthogonal projection.

Let

$$S = \frac{1}{\pi}\int\limits_{\omega} F(n) d\omega_n, \quad M = \frac{1}{2\pi}\int\limits_{\omega} L(n) d\omega_n.$$

Numbers S and M are the equal area and average integral curvature of surface B, respectively, and connected by isoperimetric inequality (Blaschke, 1916)

$$M^2 - 4\pi S \geq 4\pi^2 (R-r)^2.$$

Using this inequality and Cauchy–Bunyakowsky inequality, we have

$$\begin{aligned}
(R-r)^2 &\leq \frac{1}{4\pi^2}\left[M^2 - 4\pi S\right] = \frac{1}{4\pi^2}\left[\frac{1}{4\pi^2}\left(\int\limits_{\omega} L(n) d\omega_n\right)^2 - 4\int\limits_{\omega} F(n) d\omega_n\right] \\
&\leq \frac{1}{4\pi^2}\left[\frac{1}{4\pi^2}\int\limits_{\omega} L^2(n) d\omega_n \int\limits_{\omega} d\omega_n - \frac{1}{\pi}\int\limits_{\omega} 4\pi F(n) d\omega_n\right] \\
&\leq \frac{1}{4\pi^3}\int\limits_{\omega}\left[L^2(n) - 4\pi F(n)\right] d\omega_n \leq \frac{1}{4\pi^3}\varepsilon^2 4\pi = \frac{\varepsilon^2}{\pi^2},
\end{aligned}$$

that is $R - r \leq \dfrac{\varepsilon}{\pi}$, as stated. Notice that no regularity condition of surface B has been used in this proof. The theorem is proved.

3.4 EXISTENCE

This section is devoted to the theorems of existence of closed convex surfaces with a given functional of its orthogonal projection and illuminated parts.

Theorem 3.16. *Let $F(n)$ be a continuous, strictly positive, and even function on unit sphere ω such that $-|x|\,F\left(\frac{x}{|x|}\right)$ is conditionally positive defined on $\mathbf{R}^3$. Then there exists a unique closed convex centrally-symmetric surface B (except for a translation), for which function $F(n)$ is the area of its orthogonal projection on plane $<x,n>=0$, $n\in\omega$.*

If function $F(n)$ is k times continuously differentiable ($k\geq 4$, analytic and

$$\frac{\mathrm{d}}{\mathrm{d}t}\int\limits_{<n,p>^2>t}\frac{(\Delta+2)F(p)\,|<n,p>|}{\sqrt{<n,p>^2-t}}\mathrm{d}\omega_{\mathbf{p}}\Bigg|_{t=0}<0, \tag{3.44}$$

then surface B is at least $k-4$ times continuously differentiable.

PROOF. Let us prove the existence. By theorem (Matheron, 1975) there exists a unique symmetrical measure μ at ω such that function $F(n)$ permits representation in the form

$$F(n)=\frac{1}{2}\int\limits_{\omega}|<n,p>|\,\mu\left(\mathrm{d}\omega_{\mathbf{p}}\right). \tag{3.45}$$

Since measure μ is symmetrical on ω, and function $F(n)$ is continuous and strictly positive, then the following conditions are fulfilled:

$$\int\limits_{\omega}n\mu\left(\mathrm{d}\omega_n\right)=0,\quad \int\limits_{\omega}|<n,p>|\,\mu\left(\mathrm{d}\omega_{\mathbf{p}}\right)>0,\quad n\in\omega.$$

Hence, as it is proved in (Aleksandrov, 1937a,b; 1938a,b), there exists a unique closed convex surface B, the surface function of which is given by measure μ. Surface B is centrally-symmetric, because measure μ is symmetric. But then, by representation (3.45), $F(n)$ is the area of the orthogonal projection of this surface B on planes $<x,n>=0$.

Pass on to the proof of the regularity of surface B. Let $F(n)\in C^k(\omega)$. Consider function

$$z(n)=-\frac{1}{4\pi^2}\frac{\mathrm{d}}{\mathrm{d}t}\int\limits_{<n,p>^2>t}\frac{(\Delta+2)F(p)\,|<n,p>|}{\sqrt{<n,p>^2-t}}\mathrm{d}\omega_{\mathbf{p}}\Bigg|_{t=0}<0.$$

This function is the unique solution of

$$F(n)=\frac{1}{2}\int\limits_{\omega}|<n,p>|\,z(p)\mathrm{d}\omega_{\mathbf{p}}$$

and, by Theorem 3.6, it belongs to class $C^{k-4}(\omega)$ (and is analytic if $F(n)$ is analytic). Then with necessity function $z(n)$ is the density of measure μ, but since measure μ is a surface function, $z(n)=K^{-1}(n)$, where $K(n)$ is the Gaussian curvature of surface B. This Gaussian curvature $K(n)$ of surface B belongs to class $C^{k-4}(\omega)$ and, by condition (3.44), is strictly positive on ω. Therefore from the result of the regular solution of the Minkowski problem (Pogorelov, 1973; 1978) it follows: if $k=4$, then $K(n)$ is continuous, and the surface is smooth and of class C^1; if $k = 5$, then $K(n) \in C^1(\omega)$ and this inplicates that the surfacebelongs, at least, to class C^2, if $k = 6$, then $K(n)$ is $(k-4)$ times continuously differentiable, hence surface B is at least $(k-3)$ times continuously differentiable. Finally, if $F(n)$ is analytic, then $K(n)$ is a positive analytic function and surface B is analytic. The theorem is proved.

Theorem 3.17. *Assume a regular (k times continuously differentiable, $k \le 7$, or analytic) strictly positive even function $L(n)$ on sphere ω, satisfying condition*

$$J(n) < 0, \quad J(n) - J''_{ss}(n) \le 0, \quad n \in \omega,$$

where

$$J(n) = \frac{\mathrm{d}}{\mathrm{d}t} \int\limits_{<n,p>^2>t} \frac{(\Delta + 2)L(p)\,|<n,p>|}{\sqrt{<n,p>^2 - t}} \mathrm{d}\omega_{\mathrm{p}} \Bigg|_{t=0},$$

and index s denotes differentiation in point n along the arc of a large circle. Then there exists unique (up to translation) regular (belonging to class $C^{k-3,\alpha}(\omega)$, $0 < \alpha < 1$, or accordingly analytic) closed convex centrally-symmetric surface B, which length of boundary of the orthogonal projection on plane $<x, n> = 0$ is equal to $L(n)$, $n \in \omega$.

Proof. By its evenness and regularity function $L(n)$ admits representation in the form

$$L(n) = \frac{1}{2} \int\limits_{\omega} |<n,p>|\, \rho(p) \mathrm{d}\omega_{\mathrm{p}} \tag{3.46}$$

with a certain even, not necessarily positive, function $\rho(p)$. Equation (3.46) is uniquely solvable and function $\rho(p)$ is determined by

$$\rho(n) = -\frac{1}{4\pi^2} \frac{\mathrm{d}}{\mathrm{d}t} \int\limits_{<n,p>^2>t} \frac{(\Delta + 2)L(p)\,|<n,p>|}{\sqrt{<n,p>^2 - t}} \mathrm{d}\omega_{\mathrm{p}} \Bigg|_{t=0}.$$

By condition of Theorem 3.17 we have

$$\rho(n) \ge 0, \quad \rho(n) - \rho''_{ss}(n) \ge 0,$$

and furthermore

$$\int\limits_{\omega} n\rho(n) d\omega_n = 0.$$

since $\rho(n)$ is an even function on ω.

Thus, function $\rho(n)$ satisfies all condition of the existence theorem of a closed convex surface with given sum $\rho(n)$ of the principal radii of curvature (Pogorelov, 1973, 1978). According to this theorem there exists a unique (except for a translation) closed centrally-symmetric surface B with a given sum of the principal radii of curvature.

If $L(n) \in C^k(\omega)$, $k \ge 7$, then function $\rho(n) \in C^{k-4}(\omega)$, but then the surface of $\rho(n)$ belongs to class $C^{k-3,\alpha}(\omega)$, $0 < \alpha < 1$ (see (Pogorelov, 1973)). If $L(n)$ is analytic, then $\rho(n)$ is analytic and, hence, B is analytic (Pogorelov, 1973, 1978). The theorem is proved.

In this theorem tougher condition are set to function $L(n)$ than to function $F(n)$ in Theorem 3.16. This circumstance is the result of condition in the problem of the existence of a convex surface with a given sum of the principal radii of curvature, which are more limited with respect to the Minkowski problem.

Proof of next two theorems are conducted in a similar manner with use of the inversion formula of the integral equation (3.10), the existence theorem of a convex surface with the main function of curvature radius given and results of the regular solution of problems of Minkowski and Christoffe

Theorem 3.19. *Let a regular (k times continuously differentiable, $k \geq 4$, or analytic) function $\varphi(n)$ on unit sphere ω in $\mathbf{R}^3$ be given, satisfying conditions*

1.

$$\int_{\omega} n\varphi(n)\mathrm{d}\omega_n = 0,$$

2.

$$-\frac{1}{8\pi^2}\frac{\mathrm{d}}{\mathrm{d}t}\int_{<n,p>^2>t} \frac{(\Delta+2)\varphi_1(p)\,|<n,p>|}{\sqrt{<n,p>^2-t}}\mathrm{d}\omega_{\mathbf{p}}\Bigg|_{t=0} - \frac{1}{4\pi^2}\int_{\omega(n)} \frac{\Delta\varphi^-(p)}{<n,p>}\mathrm{d}\omega_{\mathbf{p}} > 0,$$

$$n \in \omega,$$

where

$$\varphi_1(p) = \frac{1}{a}\varphi^+(p) - \frac{b}{4\pi(a+b)}\int_{\omega}\varphi(p)\mathrm{d}\omega_{\mathbf{p}},$$

a and b are constant, $a \neq 0$, $b \neq 0$, $a+b \neq 0$. Then there exists a regular, at least $(k-3)$ times continuously differentiable (or respectively analytic) closed convex surface B, for which $\varphi(n)$ is equal to the sum of the area of its orthogonal projection on plane $<x,n>=0$ multiplied by $2a$, and the area of its illuminated part in the direction n multiplied by b. Surface B is determined uniquely, a part from a translation.

Theorem 3.20. *Let regular (k times continuously differentiable, $k \geq 7$, or analytic) function $\psi(n)$ on unit sphere ω in $\mathbf{R}^3$ be given, satisfying conditions*

1.

$$\int_{\omega} n\psi(n)\mathrm{d}\omega_n = 0,$$

2.

$$\rho(n) \leq 0, \quad \rho(n) - \rho''_{ss}(n) \leq 0, \quad n \in \omega,$$

where

$$\rho(n) = \frac{1}{8\pi^2}\frac{\mathrm{d}}{\mathrm{d}t}\int_{<n,p>^2>t} \frac{(\Delta+2)\psi_1(p)\,|<n,p>|}{\sqrt{<n,p>^2-t}}\mathrm{d}\omega_{\mathbf{p}}\Bigg|_{t=0} + \frac{1}{4\pi^2}\int_{\omega(n)} \frac{\Delta\psi^-(p)}{<n,p>}\mathrm{d}\omega_{\mathbf{p}},$$

$$\psi_1(p) = \frac{1}{a}\psi^+(p) - \frac{b}{4\pi(a+b)}\int_{\omega}\psi(p)\mathrm{d}\omega_{\mathbf{p}},$$

a and b are constant, $a \neq 0$, $b \neq 0$, $a+b \neq 0$. Index s denotes differentiation in point n on the arc of any large circle. Then there exists a unique (up to translation) regular

(belonging to class $C^{k-3,\alpha}$, $0 < \alpha < 1$ or respectively analytic) closed convex surface B, for which function $\psi(n)$ is equal to the sum of the length of the boundary of its orthogonal projection on plane $< x, n >= 0$, multiplied by $2a$, and the average integral curvature of its illuminated part in the direction n, multiplied by $2b$. Surface B is determined uniquely, except for a translation.

3.5 STABILITY

Theorems of stability definition of a convex surface by functionals of projections and illuminated parts are of interest from two points of view. From the standpoint of geometry there are very importanestimates of the proximity of surfaces, which are close to the functionals determining them. On the other hand some of the functionals appear in investigations of inverse problem of scattering, used in some approximation. In this connection theorems of stability allow us to confirm the stability definition of a scatterer by data of scattering.

At first we provide an estimate of the stability definition ofa regular convex surface by function $\varphi(n) = 2\alpha F(n) + bS(n)$.

Theorem 3.21. *If two closed, strictly convex regular surfaces B_1 and B_2 have functions $\varphi_1(n) = 2aF_1(n) + bS_1(n)$, $\varphi_2(n) = 2aF_2(n) + bS_2(n)$, $a \neq 0$, $b \neq 0$, $a + b \neq 0$ k times continuously differentiable, $k \geq 4$, and the inequality*

$$\|\varphi_2(n) - \varphi_1(n)\|_{C^4(\omega)} \leq \varepsilon, \quad \varepsilon > 0$$

is fulfilled, then at a certain location on surfaces B_1 and B_2 the difference of their support functions $H_1(n)$, $H_2(n)$ satisfies the inequality

$$\|H_2(n) - H_1(n)\|_{C(\omega)} \leq C\varepsilon^{1/5},$$

where $C > 0$ is a certain constant.

PROOF. Let $z_j(n)$ denote the product of the principal radii of curvature of surface B_j, $j = 1, 2$. Then functions $\varphi_j(n)$ can be written in the form

$$\varphi_j(n) = a \int_{\omega} |< n, p >| \, z_j(p) d\omega_p + b \int_{\omega(n)} z_j(p) d\omega_p. \tag{3.47}$$

Using inversion formula (3.17), we have

$$\begin{aligned} z_j(p) &= \frac{1}{4\pi}(\Delta + 2)\Phi_j(n) + \frac{1}{4\pi} \int_0^{\frac{\pi}{2}} \frac{(\Delta + 2)\tilde{\Phi}}{\cos\gamma} d\gamma \\ &+ \frac{1}{2\pi b}\varphi^-(n) - \frac{1}{2\pi b} \int_0^{\frac{\pi}{2}} \operatorname{tg} \gamma \tilde{\varphi}^{-\prime\prime}(\gamma, n) d\gamma, \end{aligned}$$

where

$$\Phi_j(p) = \frac{1}{a}\varphi^+(p) - \frac{b}{4\pi(a + b)} \int_{\omega} \varphi(p) d\omega_p.$$

Estimate the difference $z_2 - z_1$. By Theorem 3.6

$$\|z_2(n) - z_1(n)\|_{C(\omega)} \leq C(a,b)\, \|\varphi_2(n) - \varphi_1(n)\|_{C^4(\omega)}, \tag{3.48}$$

where $C(a,b)$ is constant, depending on a and b.

Let Q be any set of non-zero measure at ω. Since surfaces B_1 and B_2 are smooth, then it is possible to write their surface functions $G^{(j)}(Q)$ in the form

$$G^{(j)}(Q) = \int\limits_{\omega} z_j(n) \mathrm{d}\omega_n.$$

Therefore

$$\left|G^{(2)}(Q) - G^{(1)}(Q)\right| \leq \left|\int\limits_{\omega} [z_2(n) - z_1(n)]\, \mathrm{d}\omega_n\right|$$

$$\leq \operatorname{mes} Q \max_{n \in Q} |z_2(n) - z_1(n)| \leq 4\pi\, \|z_2(n) - z_1(n)\|_{C(\omega)}.$$

Hence, taking into account estimates (3.48) and $\|\varphi_2(n) - \varphi_1(n)\|_{C^4(\omega)} \leq \varepsilon$ we obtain $\left|G^{(2)}(Q) - G^{(1)}(Q)\right| \leq C_1\delta$, where $C_1 = 4\pi C(a,b)$. Using now the Volkov theorem on the stability of the solution of the Minkowski problem (Volkov, 1963), we conclude, that at a certain location on surfaces B_1 and B_2 the difference of their support functions H_1 and H_2 satisfies the inequality

$$|H_2(n) - H_1(n)| \leq C\varepsilon^{1/5},$$

with a certain constant $C > 0$. The theorem is proved.

To obtain the estimate of the stability definition of a regular convex surface by function $\psi(n) = 2aL(n) + 2bM(n)$ an estimate of the stability of the solution of the Christoffel problem is used: if sums $\rho_1(n)$ and $\rho_2(n)$ of the principal radii of curvature of closed convex surfaces B_1 and B_2 differ by not more than ε then at a certain location their support functions H_1 and H_2 differ by not more than $\left(\ln 2 - \frac{1}{4}\right)\varepsilon$. From this and the theorem of stability of the solution of the integral equation it follows

$$\psi(n) = a \int\limits_{\omega} |< n, p >| \, [R_1(p) + R_2(p)]\, \mathrm{d}\omega_{\mathrm{p}} + b \int\limits_{\omega(n)} [R_1(p) + R_2(p)]\, \mathrm{d}\omega_{\mathrm{p}}$$

at $a \neq 0$, $b \neq 0$, $a + b \neq 0$.

Theorem 3.22. *If two closed, strictly convex regular surfaces B_1 and B_2 have functions $\psi_j(n) = 2aL_j(n) + 2bM_j(n)$ k times continuously differentiable, $k \geq 4$, and inequality*

$$\|\psi_2(n) - \psi_1(n)\| \leq \varepsilon, \quad \varepsilon > 0$$

holds, then at a certain location on surfaces B_1 and B_2 the difference of their support functions H_1, H_2 satisfies inequality

$$\|H_2(n) - H_1(n)\|_{C(w)} \leq C\left(\ln 2 - \frac{1}{4}\right)\varepsilon,$$

where $C > 0$ is a certain constant.

CHAPTER 4

Integral Geometry

In this chapter the problems of integral geometry are considered. The methods and results concerning this type problems are presented. Existence and uniqueness theorems, as well as some inversion formulas are given. Some results are proved.

The problems of determing unknown objects (functions, differential forms, tensor fields) their integral known on manifolds (curves, surfaces) are called the problems of the integral geometry.

The problems of the integral geometry are connected with differential equations, inverse problems, group representations and one can find them in the geophysics, astronomy, or medicine.

The first results here were received by (Funk, 1916; Radon, 1917). They founded the integral geometry and discovered the inversion formulas.

Later, new problems of integral geometry were set, many theorems of the uniqueness and existence were proved, new inversion formulas were discovered and new applications were given.

The problems of integral geometry are studied in the works of (Blaschke, 1916; John, 1955; Gelfand et al., 1966; 1969; 1980; Gelfand and Goncharov, 1987; Kostelyanec and Reshethyak, 1954; Helgason, 1980; Semyanstyi, 1960; 1961; 1966; Kirillov, 1961; Solman, 1976; Lavrent'ev and Romanov, 1966; Lavrent'ev et al., 1970; Lavrent'ev and Bukhgeim, 1973; Lavrent'ev et al., 1986; Romanov, 1967; 1978; 1987; Lavrent'ev and Anikonov, 1967; Anikonov, 1969a; 1969b; 1978a; 1978b; 1982; 1983; Anikonov and Romanov, 1979; Anikonov and Pestov, 1990; Anikonov and Stepanov, 1991; Anikonov and Shneiberg, 1991; Bukhgeim, 1972; 1979; 1983; Mukhometov, 1977; Blagoveshchenskii, 1986; Pestov, 1985; Pestov and Sharafutdinov, 1987; Sharafutdinov, 1989; Goncharov, 1988; Palamodov and Denisjuk, 1988).

Applications and information are given in the papers by (Tikhonov et al., 1987; Natterer, 1986). One general problem of the integral geometry is to find the function

$$f(x) \in C^\infty(\mathbf{R}^n), \quad \sup_{x\in\mathbf{R}^n} (1+|x|^m) \left| \frac{\partial^\alpha f}{\partial x_1^{\alpha_1}, \ldots, \partial x_n^{\alpha_n}} \right| < \infty, \quad \forall m \geq 0$$

if we know the function

$$\hat{f}(y) = \int\limits_{B(y)} f(x) \mathrm{d}\mu, \quad y \in \mathbf{R}^n,$$

where $B(y) \subset \mathbf{R}^n$ is a smooth manifold, $y \in \mathbf{R}^n$, and μ is a measure on $B(y)$ (Gelfand et al., 1966). The Radon transform is

$$\hat{f}(\omega,p) = \int\limits_{(\omega,x)=p} f(x)\mathrm{d}\mu, \quad y = (\omega,p),$$

$$B(y) = \{x : (\omega,x) = p\}, \quad \omega \in \mathbf{R}^n, \quad |\omega| = 1, \quad p \in \mathbf{R}, \quad x \in \mathbf{R}^n.$$

In this chapter we shall discuss some methods of the integral geometry, especially in the curve case and some applications. At first we give some review.

4.1 INVERSION FORMULAS

4.1.1. The Radon formulas

The inversion formulas in problems of the integral geometry play a very important role in the theory and applications.

Theorem 4.1. *(Radon, 1917; Gelfand et al., 1966; Helgason, 1980). Let a function $\hat{f}(\omega,p)$ be the Radon transform*

$$\hat{f}(\omega,p) = \int\limits_{(x,\omega)=p} f(x)\mathrm{d}\nu.$$

Then the function $\hat{f}(\omega,p)$, $\omega \in \Omega$, $\Omega = \{\omega, |\omega| = 1\}$, $p \in \mathbf{R}$, satisfies the conditions:

1. $\hat{f}(\omega,p) \in C^\infty(\Omega \times \mathbf{R})$,

2.

$$\sup_{p\in\mathbf{R}}(1 + |p|^m)\left|\frac{\partial^\alpha \hat{f}}{\partial p^\alpha}\right|, \quad \forall m \geq 0, \quad \alpha \geq 0,$$

3. $\int\limits_{\mathbf{R}} \hat{f}(\omega,p)p^k\mathrm{d}p$ *is a homogeneous polynomial of the degree k depending on $\omega = (\omega_1,\ldots,\omega_n)$ and the inversion Radon formula*

$$f(x) = \frac{(-4\pi)^{(n-1)/2}\Gamma(\frac{n}{2})}{\Gamma(\frac{1}{2})}\Delta^{\frac{n-1}{2}} \int\limits_{\Omega} \hat{f}(\omega,(\omega,x))\mathrm{d}\omega$$

holds, where $\Delta^{\frac{n-1}{2}}$ is the Laplace operator of degree $(n-1)/2$.

Other similar formulas are contained in the papers by (Semyanistyi, 1960; 1961; 1966; Gelfand et al., 1980).

If $n = 2$, $x_1 = x$, $x_2 = y$, $x^2 + y^2 \leq 1$

$$\hat{f}(\varphi,p) = \int\limits_{x\cos\varphi + y\sin\varphi = p} f(x,y)\mathrm{d}l, \quad 0 \leq \varphi \leq 2\pi, \quad -1 \leq p \leq 1,$$

then the different inversion formulas may be written on the basis of Theorem 4.1.

$$
\begin{aligned}
f(x,y) &= \frac{1}{(2\pi)^2}\Delta \int_0^{\pi} \mathrm{d}\varphi \int_{-1}^{1} \hat{f}(\varphi,p)\ln|p-rn|\,\mathrm{d}p \\
&= \frac{1}{(2\pi)^2}\int_0^{\pi} \mathrm{d}\varphi \int_{-1}^{1} \frac{\partial^2 \hat{f}(\varphi,p)}{\partial p^2}\ln|p-rn|\,\mathrm{d}p \\
&= -\frac{1}{(2\pi)^2}\int_0^{\pi} \mathrm{d}\varphi \int_{-1}^{1} \frac{\partial \hat{f}(\varphi,p)}{\partial p}\frac{1}{|p-rn|}\mathrm{d}p \\
&= \frac{1}{(2\pi)^2}\int_0^{\pi} \mathrm{d}\varphi \int_{-1}^{1} \hat{f}(\varphi,p)\frac{1}{(p-rn)^2}\mathrm{d}p \\
&= \frac{1}{(4\pi)^2}\int_{-\infty}^{\infty}\int_{-\infty}^{\infty}\left[\int_{-1}^{1}\hat{f}\left(\arctan\frac{\omega_1}{\omega_2},t\right)\mathrm{e}^{\mathrm{i}t\sqrt{\omega_1^2+\omega_2^2}}\mathrm{d}t\right]\mathrm{e}^{\mathrm{i}\omega_1 x+\mathrm{i}\omega_2 y}\mathrm{d}\omega_1\mathrm{d}\omega_2,
\end{aligned}
$$

$$rn = x\cos\varphi + y\sin\varphi.$$

4.1.2. The other formulas

In this section we consider the problem of the integral geometry to find the function $f(x)$, $x \in \mathbf{R}^3$ such that

$$\hat{f}(\gamma) = \int_{\gamma} f(x)\mathrm{d}l,$$

where γ is a line passing through a smooth curve $\Gamma \subset \mathbf{R}^3$. The function $f(\gamma)$ is supposed to be known for all lines such that $\gamma \cap \Gamma \neq \emptyset$.

Theorem 4.2. *(Kirillov, 1961) A function*

$$f(x) \in C^{\infty}(\mathbf{R}^3), \quad \sup_{x\in\mathbf{R}^3}(1+|x|^m)\left|\frac{\partial^{\alpha} f}{\partial x_1^{\alpha_1},\dots,\partial x_n^{\alpha_n}}\right| < \infty, \quad \forall m \geq 0, \quad \alpha \geq 0$$

is restored by $\hat{f}(\gamma)$ and the inversion formulas hold if and only if $\{x : (x,\omega) = p\} \cap \Gamma \neq \emptyset$ for almost all (ω,p), $|\omega| = 1$, $p \in \mathbf{R}$.

(Tuy, 1983) discovered the inversion formulas for the functions with compact support when the curve is bounded, provided that almost every hyperplane intersecting the support of the unknown function meets the curve Γ transversally in some point. (Finch, 1985) investigated the analogous inversion procedure for the class of the curves which did not satisfy the hypotheses of Tuy.

(Blagoveshchenskii, 1986) found the inversion formulas when the curve Γ is a circle in $\mathbf{R}^3$. He found essentially new formulas. The uniqueness theorems for the integral geometry problems when unknown functions have a compact support were studied by (Anikonov, 1978; Helgason, 1980; Leahy et al., 1979).

4.1.3. Integral geometry on the sphere

The problems of integral geometry on a sphere were considered in the papers of (Blaschke, 1916; Semyanistyi, 1960; 1961; 1966; Helgason, 1980). The unknown functions were supposed to be even on the sphere.

Let n, p be a point of the unit sphere S in $\mathbf{R}^3$ with the center at the origin, (n,p) the inner product $n \in S$, $p \in S$, $\Gamma(n) = \{p \in S, (n,p) = 0\}$, $S(n) = \{p \in S, (n,p) \geq 0\}$, Δ be the Laplace-Beltrami operator, and a, b be the constants.

One of the integral geometry problems is to determine the function $f(n) \in C^\infty(S)$, provided the function $\hat{f}(p)$ is known. $p \in S$,

$$\hat{f} = 2a \int_{\Gamma(p)} f(n) \mathrm{d}\Gamma + b \int_{S(n)} (\Delta + 2) f(n) \mathrm{d}S,$$

the function $f(p)$ is not necessary even, (see Section 3.2).

Theorem 4.3. *(Anikonov and Stepanov, 1991) Let $a \neq 0$, $b \neq 0$, $a + b \neq 0$. The inversion formula holds*

$$\hat{f}(n) = \frac{1}{4\pi} \int_S \Phi(p)(1 - (n,p)) \ln(1 - (n,p)) \mathrm{d}S,$$

with

$$\Phi(n) = -\frac{1}{8\pi^2} \frac{\mathrm{d}}{\mathrm{d}t} \int_{(n,p)^2 > t} \frac{(\Delta + 2)\varphi_1(p)(n,p) \mathrm{d}S}{\sqrt{(n,p)^2 - t}} \Bigg|_{t=0} - \frac{1}{4\pi^2 b} \int_{S(n)} \frac{\Delta \varphi^-(p) \mathrm{d}S}{(n,p)}$$

$$\varphi_1 = \frac{1}{a}\varphi^+ - \frac{b}{4\pi a(a+b)} \int_S \varphi \mathrm{d}S, \quad \varphi^\pm = \frac{\varphi(p) \pm \varphi(-p)}{2}.$$

4.1.4. Equation for function $\mu(\varphi)$

In this section we consider the generalized problem of the integral geometry, namely, we seek two functions $f(x) \in C^\infty(|x| \leq 1)$, $x = (x_1, x_2)$ and $\mu(\varphi) \in C^\infty([0, 2\pi])$, given the function

$$\hat{f}(\varphi, p) = \int_{x \cos\varphi + y \sin\varphi = p} f(x) \, \mathrm{e}^{\mu(\varphi)(-x_1 \sin\varphi + x_2 \cos\varphi)} \mathrm{d}l.$$

It should be noted that if the function $\mu(\varphi)$ is known, the problem is linear and the inverse formula for the function $f(x)$ holds (Kuchment, 1989).

Theorem 4.4. *(Anikonov and Shneiberg, 1991) The function $\mu(\varphi)$, $0 \leq \varphi \leq 2\pi$ is the solution of the differential equation*

$$\int_{-1}^{1} \left[\frac{\partial \hat{f}(\varphi, p)}{\partial \varphi} + \mu(\varphi) p \hat{f}(\varphi, p) \right] e^{\mu'(\varphi) p} \mathrm{d}p = 0, \quad \frac{\mathrm{d}\mu}{\mathrm{d}\varphi} = \mu'.$$

Other similar results are contained in the papers of (Anikonov, 1983; Anikonov and Shneiberg, 1991).

In particular, the uniqueness class of two-dimensional integral geometry problem consists of the smooth functions $f(x_1, x_2, \varphi)$ such that

$$\left(\frac{\partial}{\partial x_1}\sin\varphi - \frac{\partial}{\partial x_2}\cos\varphi\right)\frac{\partial f}{\partial \varphi} = 0.$$

4.1.5. Support theorems

Let $f(P)$ be the Radon transform

$$\hat{f}(P) = \int\limits_P f(x)\mathrm{d}\mu,$$

where P is a plane.

Theorem 4.5. *(Helgason, 1980). Let the following conditions hold*

1. $f(x) \in C^\infty(\mathbf{R}^n)$,

2. $|x^n|\,|f(x)| \leq \infty$, $\forall k \geq 0$, $x \in \mathbf{R}^n$,

3. $\exists A > 0$ *such that* $\hat{f}(P) = 0$ *for all* $P : \rho(0, P) \geq A$, *where* ρ *is a distance in* $\mathbf{R}^n$.

Then $f(x) = 0$, $|x| > A$.

Theorem 4.6. *(Logvinenko, 1988). Let the following conditions hold*

1. $f(x) \in C^\infty(\mathbf{R}^n)$, $f(x) = o(|x|^{-k})$, $\forall k > 0$, $x \to \infty$,

2. *the function*

$$\hat{f}(\omega, p) = \int\limits_{(\omega,x)=p} f(x)\mathrm{d}\mu$$

satisfies the condition $\left|\hat{f}\right| \leq c_\omega\, \mathrm{e}^{-\varepsilon_\omega p}$, *and* $c_\omega > 0$, $\varepsilon_\omega > 0$ *are constants,*

3. $\exists A > 0$ *and a set* $e \in \Omega$, $\mu e \neq 0$ *such that* $\hat{f}(\omega, p) = 0$, $\omega \in e$, $|p| > A$.

Then Support $f(x)$ *is a compact set.*

4.1.6. Inversion formulas for tensor fields

Let $f(x,\xi) = \sum\limits_{i=0}^{m} f_{i_1\ldots i_m}\xi_{i_1}\ldots\xi_{i_m}$, $x \in \mathbf{R}^n$, $|\xi| = 1$, $m \geq 0$ be a tensor field such that

1. $f(x,\xi) \in C^\infty(\mathbf{R}^n \times |\xi| = 1)$,

2. $\sup\limits_{x\in\mathbf{R}^n}(1 + x^k)\left|\dfrac{\partial^\alpha f}{\partial x_1^{\alpha_1}, \ldots \partial x_n^{\alpha_n}}\right| < \infty, \quad \forall k \geq 0, \quad \alpha_i \geq 0.$

We consider here the problem of the integral geometry to find $f(x,\xi)$ if we know the function

$$\hat{f}(x,\xi) = \int_{-\infty}^{\infty} f(x+t\xi,\xi)dt.$$

Theorem 4.7. *(Sharafutdinov, 1989) If the conditions 1 and 2 hold, then*

3. $f = \tilde{f} + \mathbf{d}v$, $\delta\tilde{f} = 0$,

4.

$$\tilde{f} = (-\Delta)^{1/2}\left[\sum_{k=0}^{[m/2]} c_k(\mathbf{i} - \Delta^{-1}\mathbf{d}^2)^k \mathbf{j}^k\right]\mu^m \hat{f}.$$

c_k are constants, $\mathbf{i}$, Δ, $\mathbf{d}$, $\mathbf{j}$, μ are the operators defined in the paper of (Sharafutdinov, 1989).

4.2 THE UNIQUENESS AND SOLVABILITY

4.2.1. Integral geometry problem uniqueness of the solution

Consider the two-dimensional problem of finding a smooth function $f(r,\varphi)$, $0 \le r \le 1$, $0 \le \varphi \le 2\pi$ from the function

$$\hat{f}(\rho,\alpha) = \sum_{j=1}^{2}\int_{\rho} f(r,\varphi_j)\sqrt{1+\left(r\frac{\partial\varphi_j}{\partial r}\right)^2}\,dr, \quad 0 \le \rho \le 1, \quad 0 \le \alpha \le 2\pi,$$

$$\varphi_j = \alpha - (-1)^{-j}\sqrt{r-\rho\psi_j(r,\rho)}, \quad \psi \in C^1(0<\rho\le r\le 1), \quad \psi_1(\rho,\rho) = \psi_2(\rho,\rho).$$

This problem is equivalent to the problem in the integral geometry where curves have group properties (Romanov, 1967).

Theorem 4.8. *(Romanov, 1967) If $\hat{f}(\rho,\alpha) = 0$, $0 \le \rho \le 1$, $0 \le \alpha \le 2\pi$, then $f(r,\varphi) = 0$, $0 \le r \le 1$, $0 \le \varphi \le 2\pi$.*

The generalizations and applications of this result may be found in the book by (Romanov, 1987). In particular, using the inversion formulas one can prove the uniqueness theorem when manifolds are not a plane or a sphere.

4.2.2. Uniqueness of the solution of the integral geometry problems in the class of analytic functions

Let the smooth manifolds $B(z,p)$, $z \in \mathbf{R}^n$, $p > 0$ satisfy the condition

$$B(z,p) \subset \Omega^+(z,p) = \{(x,y) : x \in \mathbf{R}^n, \quad y > 0, \quad |x-z|^2 + y^2 < p, \quad n \ge 1\}.$$

Theorem 4.9. *(Anikonov, 1978) Let the function* $f(x,y) \in C^{\infty}(x \in \mathbf{R}^n, y \geq 0)$ *be analytic in the variable* y *in the domain* $x \in \mathbf{R}^n$, $y \geq 0$. *If*

$$\int_{B(z,p)} f(x,y)\mathrm{d}\mu = 0, \quad z \in \mathbf{R}^n, \quad p > 0,$$

then $f(x,y) = 0$, $x \in \mathbf{R}^n$, $y \geq 0$.

4.2.3. Differential identities and integral geometry

The differential identities can be used well to prove the uniqueness, stability, and existence theorems of the solution in integral geometry. Here we consider only the two-dimensional case. Let $\mathbf{M}$, $\tilde{\mathbf{M}}$ be the differential operators (see Section 2.9)

$$\mathbf{M}W = \left(\frac{\partial W}{\partial x} - c\right)\cos\theta + \left(\frac{\partial W}{\partial y} - d\right)\sin\theta,$$

$$\tilde{\mathbf{M}}W = \left(\frac{\partial W}{\partial x} - c\right)\sin\theta - \left(\frac{\partial W}{\partial y} - d\right)\cos\theta,$$

where $W(x,y,z)$, $\theta(x,y,z)$, $c(x,y)$ $(\partial c/\partial z = 0)$, $d(x,y)$ $(\partial d/\partial z = 0)$ are functions in C^2 in the domain $x^2 + y^2 \leq 1$, $0 \leq z \leq 2\pi$. Here we give the application of Theorem 2.17 when D is a circle.

Theorem 4.10. *(Anikonov, 1978) The following identity holds*

$$\tilde{\mathbf{M}}W\frac{\partial}{\partial z}\mathbf{M}W - \mathbf{M}W\frac{\partial}{\partial z}\tilde{\mathbf{M}}W = -\frac{\partial\theta}{\partial z}\left[\left(\frac{\partial W}{\partial x} - c\right)^2 - \left(\frac{\partial W}{\partial y} - d\right)^2\right]$$

$$+\frac{\partial}{\partial y}\left(\frac{\partial W}{\partial x}\frac{\partial W}{\partial z}\right) - \frac{\partial}{\partial x}\left(\frac{\partial W}{\partial y}\frac{\partial W}{\partial z}\right) - \frac{\partial}{\partial z}\left(c\frac{\partial W}{\partial y} - d\frac{\partial W}{\partial x}\right),$$

$$x^2 + y^2 \leq 1, \quad 0 \leq z \leq 2\pi.$$

In the case when $c = 0$ and $d = 0$ the identity is the same as obtained by (Mukhometov, 1977). Let $\lambda(x,y)$, $\mu(x,y)$, $a(x,y)$, $b(x,y)$, $c(x,y)$, and $d(x,y)$ be smooth functions in the domain $x^2 + y^2 \leq 1$, and

$$\gamma(z,t) = \{x(s;z,t), \quad y(s;z,t), \quad x|_{s=0} = \cos z, \quad y|_{s=0} = \sin z\}$$

be a set of smooth extremals of the functional

$$\int_{\gamma}\left[a\mathrm{d}x + b\mathrm{d}y + \lambda\sqrt{\mathrm{d}x^2 + \mathrm{d}y^2}\right],$$

which joins the points $(\cos z, \sin z)$ and $(\cos t, \sin t)$. Assume that the set γ is regular. In this section we consider the non-linear problem of integral geometry to find the functions λ, μ, a, b, c, and d if the functions

$$f(z,t) = \int_{\gamma(z,t)}\left[c\mathrm{d}x + d\mathrm{d}y + \mu\sqrt{\mathrm{d}x^2 + \mathrm{d}y^2}\right],$$

$$\theta(z,t) = \arctan \frac{y'}{x'}\bigg|_{s=0}$$

are known in the domain $0 \le z \le 2\pi$, $0 \le t \le 2\pi$. From Theorem 4.10 it follows.

Theorem 4.11. *(Anikonov, 1978) Let the functions λ, μ be known for (x,y) with $x^2+y^2=1$. Then the functions $\lambda(x,y)$, $\mu(x,y)$, $\frac{\partial a}{\partial y}-\frac{\partial b}{\partial x}$, and $\frac{\partial c}{\partial y}-\frac{\partial d}{\partial x}$ are uniquely determined from the functions $f(z,t)$ and $\theta(z,t)$.*

The problems in multidimensional cases were considered by (Romanov, 1978; 1987) and (Pestov, 1985). Pestov was the first to use the curvature of Riemann spaces in identities.

4.2.4. Integral geometry and evolution equations

In the case of a line integral the integral geometry problems reduce to Cauchy problems for evolution equations (Anikonov, 1969a; Lavrent'ev and Anikonov, 1967). After solving these problems, the unknown functions can be found by formulas.

The two-dimensional problem of integral geometry reduced to the inverse problem for the transport equation:

$$\frac{\partial W}{\partial x}\cos\varphi + \frac{\partial W}{\partial y}\sin\varphi + K(x,y,\varphi)\frac{\partial W}{\partial \varphi} = f(x,y),$$

$$W\big|_{y=0} = \hat{f}(x,\varphi), \quad x \in \mathbf{R}, \quad 0 \le \varphi \le \pi, \quad y \ge 0,$$

where the functions $W(x,y,\varphi)$ and $f(x,y)$ are unknown. Information is given by the function $\hat{f}(x,\varphi)$. The function $K(x,y,\varphi)$ is the curvature.

Theorem 4.12. *(Anikonov, 1978a) Let a smooth function $u(x,y,z)$ be the solution to the Cauchy problem*

$$\frac{\partial u}{\partial y} = -\int_0^1 \frac{\partial}{\partial p}\left[\frac{\partial u(x,y,p)}{\partial x}\sqrt{1-p^2} + K\frac{\partial u(x,y,p)}{\partial p}(1-p^2)\right]_{p=z\eta} d\eta \equiv Au,$$

$$u\big|_{y=0} = \hat{f}(x, \arcsin z).$$

Then

$$f(x,y) = \left[\frac{\partial u}{\partial x} + K(x,y,0)\frac{\partial u}{\partial z}\right]_{z=0}.$$

REMARK . If the function K does not depend on y, we have

$$u(x,y,z) = e^{yA}\hat{f}.$$

4.2.5. On the solvability of a problem of integral geometry.

We consider here the question of solvability of a integral geometry problem in the case when the integration curves are geodesics of a fixed analytic metrics. Let

$$ds^2 = B(x,y)(dx^2 + dy^2)) \tag{4.1}$$

be an analytic metrics defined on the $(x,y)-$ plane. Henceforth, we suppose that $B > 0$ and $\partial B/\partial y < 0$. We denote by $\tilde{\gamma}(\xi,\varphi)$ the geodesic of the metrics (4.1) emanating from the point $(\xi, 0)$ at an angle φ. Let $\gamma(\xi,\varphi)$ be the part of $\tilde{\gamma}(\xi,\varphi)$ lying in the half-plane $y \geq 0$. The problem we consider is to find the function $f(x,y)$ in the region $y \geq 0$ if the function

$$\hat{f}(\xi,\varphi) = \int\limits_{\gamma(\xi,\varphi)} f(x,y)\sqrt{dx^2 + dy^2}$$

is given for $-\delta \leq \xi \leq \delta$ and $0 \leq \varphi \leq \alpha$.

We start with giving preliminaries. Denote by $\eta(\xi,\varphi)$, $\eta > \xi$, a point of the line $y = 0$ that belongs to the geodesic $\gamma(\xi,\varphi)$, and let $\theta(\xi,\varphi)$ be the angle between $\gamma(\xi,\varphi)$ and the x - axis at the point $(\eta, 0)$. Set $a(\xi,\varphi) = B(\xi,0)\sin\varphi$, $b(\xi,\varphi) = B(\eta(\xi,\varphi),0)\sin\varphi$. Assuming

$$t = \int\limits_{\gamma} \frac{\sqrt{dx^2 + dy^2}}{B},$$

the equations for geodesics of the metrics (4.1) become

$$\frac{d^2x}{dt^2} = B\frac{\partial B}{\partial x}, \quad \frac{d^2y}{dt^2} = B\frac{\partial B}{\partial y}, \quad x'^2 + y'^2 = B^2. \tag{4.2}$$

It is easily seen that any first analytic integral of (4.2) can be written in the form

$$\tilde{r} = u(x,y,y')x' + v(x,y,y')$$

where $u(x,y,y)$ is an even analytic function of y', and $v(x,y,y')$ is an odd analytic function of y'.

In the following we shall consider real analytic functions in a neighbourhood of the origin of the real Euclidean space $\mathbf{R}^2$.

Let E be the set of all analytic functions $g(x,z)$, $x \in \mathbf{R}$, $z \in \mathbf{R}$. We denote by T the set of analytic functions $r(x,z)$ generated by the first integrals of system (4.2); namely, if

$$\tilde{r} = u(x,y,y')x' + v(x,y,y')$$

is a first integral of (4.2), then

$$r(x,z) = u(x,0,z) + v(x,0,z).$$

Let E/T be the quotient space of the equivalence classes: $g_1 \sim g_2$, $g_1 - g_2 \in T$. Let A be the set of functions $f(\xi,\varphi)$, with

$$\begin{aligned}\hat{f}(\xi,\varphi) &= \frac{1}{2}\,[g(\eta,b)(1 + B(\eta,0)\cos\varphi) - g(\eta,-b)(1 - B(\eta,0)\cos\theta)] \\ &\quad -\frac{1}{2}\,[g(\xi,a)(1 + B(\xi,0)\cos\varphi) - g(\xi,-a)(1 - B(\xi,0))\cos\varphi],\end{aligned}$$

where η, b, a, θ are the above defined functions, $g \in E$. Denote such a representation of the functions $f(\xi,\varphi)$ by Ag and write $\hat{f} = Ag$.

We denote by Dg, $g \in E$, the analytic function $f(x,y)$ of the variables (x,y) defined as follows: we consider the system of the equations

$$\frac{\partial u_k}{\partial y} = -\frac{\partial v_k}{\partial x} - 2(k+1)B\frac{\partial B}{\partial y}u_{k+1},$$

$$\frac{\partial v_k}{\partial y} = \frac{\partial u_k}{\partial x} - u_{k+1}B\frac{\partial B}{\partial x} - \frac{\partial u_{k+1}}{\partial x}B^2 - (2k+3)v_k\frac{\partial B}{\partial y},$$

with the Cauchy data

$$u_k\big|_{y=0} = u_k^0(x), \quad v_k\big|_{y=0} = v_k^0(x),$$

such that

$$\sum_{k=0}^{\infty}\left[\left(u_k^0 z^{2k} + v_k^0 z^{2k+1}\right)\right] = g(x,z), \quad g \in E.$$

If this problem has an analytic solution u_k, v_k, $k = 1,2,\ldots$, then by definition

$$Dg = u_0\frac{\partial B}{\partial x} + v_0\frac{\partial B}{\partial y} + \frac{\partial u_0}{\partial x}B.$$

Theorem 4.13. *(Anikonov ,1978a)*

1. *A necessary and sufficient condition for the solvability of the integral geometry problem in the class of analytic functions $f(x,y)$ is that the function*

$$\hat{f}(\xi,\varphi) = \int_{\gamma(\xi,\varphi)} f(x,y)\sqrt{\mathrm{d}x^2 + \mathrm{d}y^2}$$

can be represented in the form $\hat{f} = Ag$. Moreover, for any function $g(x,z)$ the operation Dg is defined, and

$$Ag = \int_{\gamma(\xi,\varphi)} Dg\sqrt{\mathrm{d}x^2 + \mathrm{d}y^2}.$$

2. *$Ag_1 = Ag_2$ if and only if the equivalence classes of E/T corresponding to the functions g_1 and g_2 coincide.*

3. *If*

$$\int_{\gamma} f_1\sqrt{\mathrm{d}x^2 + \mathrm{d}y^2} = \int_{\gamma} f_2\sqrt{\mathrm{d}x^2 + \mathrm{d}y^2}$$

and if f_i, $i = 1,2$, are analytic functions, then $f_1(x,y) = f_2(x,y)$.

We shall give the proof of this theorem in Section 4.5.

4.2.6. Integral geometry and the structure of Riemann spaces

Let M be an n - dimensional ($n \geq 2$) simply-connected compact Riemann manifold of the class C^∞ with a metric tensor $\mathbf{g} = (g_{ij})$ and a strictly convex boundary ∂M.

We introduce the following notation: $\mathbf{R}_{ijkl}$ is the curvature tensor of the metric $\mathbf{g}$, T_x is the tangent space at x, ΩM is the tangent bundle of unit vectors, $\Omega M = \{(x,\xi) : x \in M, \quad \xi \in T_x\}$,

$$|\xi| = \sqrt{g_{ij}(x)\xi^i\xi^j}$$

(here and below summation from 1 to n over repeated subscripts and supscripts is understood),

$$\Omega\partial M = \{(x,\xi) : x \in \partial M, \quad \xi \in T_x, \quad |\xi| = 1, \quad (\xi, v(\xi)) \leq 0\},$$

where $v(x)$ is the outward normal to ∂M at the point x, and $\gamma_{x,\xi}(t)$ is the geodesic defined by the initial data $\gamma_{x,\xi}(0) = x$, $\dot{\gamma}_{x,\xi}(0) = \xi$.

The set of the smooth (class C^∞) covariant symmetric tensor fields of valency m, $m \geq j$ on M and ∂M is denoted by $S_m M$ and $S_m \partial M$, respectively. Define the operation of symmetric covariant differentiation $\mathrm{d}S_m M \to S_{m+j}M$, $\mathrm{d} = \sigma\nabla$, where σ is the symmetric operator and ∇ is the operator of the covariant differentiation with respect to the metric $\mathbf{g}$. Henceforth we shall assume that M is the dispersing manifold: any geodesic of it leaves a compactum $K \subset M$, $K \cap \partial M = \emptyset$. We define the function $t^0(x,\xi)$, $(x,\xi) \in \Omega^- M$ as the length of the geodesic $\gamma_{x,\xi}$, $(x,\xi) \in \Omega^- M$, and call it the hodograph of the metric $\mathbf{g}$. By the assumptions we have made about M, the hodograph $t^0(x,\xi)$ is well-defined on $\Omega^- M$.

Suppose that $f = f_{i_1\ldots i_m}$ and $f \in S_m M$.

Consider the integrals along the geodesic joining points of ∂M

$$f(x,\xi) = \int_0^{t^0(x,\xi)} f_{i_1\ldots i_m}(\gamma_{x,\xi}(t))\dot{\gamma}^{i_1}_{x,\xi}\cdots\dot{\gamma}^{i_m}_{x,\xi}\mathrm{d}t. \tag{4.3}$$

Theorem 4.14. *(Pestov and Sharafutdinov, 1987) Let M be a compact dispersing manifold with a strictly convex boundary ∂M, and suppose that the sectional curvatures of M are nonpositive, that is at each point $x \in M$, $\mathbf{R}_{ijkl}\xi^i\eta^j\xi^k\eta^l \leq 0$ for any $\xi, \eta \in T_x$. Then the integral (4.3) vanishes when $(x,\xi) \in \Omega^-\partial M$ if and only if*

$$f = dv, \quad v \in S_{m-1}M, \quad v|_{\partial M} = 0, \quad m \geq 1, \quad f = 0, \quad m = 0.$$

It turns out that such theorems of the integral geometry enable us to determine the structure of the Riemann space from the integral information.

Theorem 4.15. *(Anikonov and Pestov, 1989) Suppose that the conditions of Theorem 4.14 are satisfied. Then the equations of geodesics admit a first integral*

$$J(\gamma_{x,\xi}(t), \dot{\gamma}_{x,\xi}(t)) = u_{i_1\ldots i_m}\dot{\gamma}^{i_1}_{x,\xi}\cdots\dot{\gamma}^{i_m}_{x,\xi},$$

where $u = (u_1, u_2, \ldots, u_m) \in S_m M$ if and only if there is a field $u \in S_m \partial M$ such that

$$u^0_{i_1\ldots i_m}(x)\xi^{i_1}\ldots\xi^{i_m} = u^0_{i_1\ldots i_m}(\gamma_{x,\xi}(t^0(x,\xi)))\dot{\gamma}^{i_1}_{x,\xi}(t^0(x,\xi))\ldots\dot{\gamma}^{i_m}_{x,\xi}(t^0(x,\xi)), \quad (x,\xi) \in \Omega\partial M.$$

We shall give detailed analysis of this results in Section 4.4.

4.3 SOME APPLICATIONS

We consider a multidimensional inverse kinematic problem and obtain the necessary and sufficient conditions in order that some differential relations for unknown functions are held.

The proof is based on the theorems of the integral geometry. One formulation of the multidimensional inverse kinematic problem consists in the following: in the ball $\omega = \{x : x \leq 1\}$ of the real Euclidean space of variables $x = (x_1, \dots, x_n)$, $n \geq 2$ we consider a 3 times continuously differentiable metric

$$ds^2 = \lambda(x)^2 \, |dx|^2 , \tag{4.4}$$

such that any two points $p \in \omega$, $q \in \omega$ can be joined by a unique geodesic $\gamma(p,q)$ of the metric (4.4); the function

$$W(p,q) = \int\limits_{\gamma(p,q)} \lambda(x) \, |dx| , \quad |p| = 1, \quad |q| = 1,$$

is assumed to be known, and it is required to find $\lambda(x)$ for $|x| \leq 1$.

This problem has an important physical interpretation in the seismology. The function $\lambda(x) = 1/v(x)$ is the inverse to the propagation speed of a perturbation. The function $W(p,q)$ is the time this propagation takes along the geodesic $\gamma(p,q)$. The formulated problem consists in determining the velocity $v(x)$ on the basis of the times $W(p,q)$, $|p| = 1$, $|q| = 1$.

Some results on the uniqueness, stability and existence of the multidimensional kinematic problem solutions and of the integral geometry closely related to the inverse kinematic problem are contained in the papers of (Anikonov, 1978a; b; Romanov, 1987). The constructive techniques to solve these problems are particularly interesting in some applications. Here, necessary and sufficient conditions are presented so that the differential relations for the known function $W(p,q)$, $|p| = 1$, $|q| = 1$ and the unknown function $\lambda(x)$, $|x| \leq 1$ are simultaneously satisfied. We proceed to formulate the results. Suppose that the components $u(x), k = 1, 2, \dots, n$ of the vector-valued function $u(x)$ satisfy the system of equations

$$\frac{\partial u_k}{\partial x_j} = -\frac{\partial u_j}{\partial x_k}, \quad k \neq j, \quad \frac{\partial u_k}{\partial x_k} = \frac{\partial u_j}{\partial x_j} \tag{4.5}$$

In the case for $n = 2$ this is the system of the Cauchy-Riemann equations; for $n > 2$ its general solution is given by

$$u(x) = \mathbf{A}x - Bx^2 + 2x(B,x) + C$$

where $\mathbf{A}$ is a constant matrix with the elements A_{jk} such that $A_{jk} = -A_{kj}$, $k \neq j$, $A_{jj} = A_0$, B and C are constant vectors, and (B,x) is an inner product. There is a considerable difference between the classes of the vector-valued functions $u(x)$ for $n = 2$ and $n > 2$.

Let $u_k(x) = (u_{k_1}, \dots, u_{k_n})$ and $v(x) = (v_{k_1}, \dots, v_{k_n})$ be vector-valued functions satisfying (4.5) for any k. We set

$$a_{ij} = \frac{1}{2} \sum_{k=1}^{n} \left(u_{k_j} v_{k_i} + u_{k_i} v_{k_j} \right) ,$$

and let

$$Q = \sum_{i,j=1}^{n} a_{ij} x'_i x'_j.$$

Along any geodesic γ we have

$$\frac{dQ}{dt} = \sum_{i=1}^{n} \left(\sum_{k=1}^{n} \left(v_{k_i} \tilde{\mathbf{L}}_k^u \lambda + u_{k_i} \tilde{\mathbf{L}}_k^v \lambda \right) \right) x', \quad |x'| = \lambda,$$

where

$$\tilde{\mathbf{L}}_k^u \lambda = \lambda(x) \, (\operatorname{grad} \lambda, u_k) + \frac{\lambda^2}{n} \sum_{i=1}^{n} \frac{\partial u_{k_i}}{\partial x_i},$$

$$\tilde{\mathbf{L}}_k^v \lambda = \lambda(x) \, (\operatorname{grad} \lambda, v_k) + \frac{\lambda^2}{n} \sum_{i=1}^{n} \frac{\partial v_{k_i}}{\partial x_i}$$

Consider the differential expression

$$\mathbf{L}w = \sum_{k=1}^{n} \left[\left(u_k(q), \frac{\partial w}{\partial q} \right) \left(v_k(q), \frac{\partial w}{\partial q} \right) - \left(u_k(p), \frac{\partial w}{\partial p} \right) \left(v_k(p), \frac{\partial w}{\partial p} \right) \right],$$

where

$$w(p,q) = \int_{\gamma(p,q)} \lambda(x) \, |dx| \,, \quad \frac{\partial w}{\partial q} = \left(\frac{\partial w}{\partial q_1}, \ldots, \frac{\partial w}{\partial q_n} \right), \quad \frac{\partial w}{\partial p} = \left(\frac{\partial w}{\partial p_1}, \ldots, \frac{\partial w}{\partial p_n} \right)$$

and $u(x)$ and $v(x)$ are vector-valued functions satisfying (4.5).

Theorem 4.16. *(Anikonov and Pestov, 1990a) In order that*

$$\mathbf{L}w = \varphi(q) - \varphi(p), \quad |q| = 1, \quad |p| = 1$$

for some continuously differentiable function $\varphi(x)$, $|x| \le 1$ it is necessary and sufficient that

$$\frac{\partial \varphi}{\partial x_i} = \sum_{k=1}^{n} \left(v_{k_i} \tilde{\mathbf{L}}_k^u \lambda + u_{k_i} \tilde{\mathbf{L}}_k^v \lambda \right), \quad i = 1, 2, \ldots, n.$$

To formulate another result in the inverse kinematic problem we suppose that $u(x)$ satisfies (4.5) and has no more restrictions. Consider the differential operators $\mathbf{L}w$, $\tilde{\mathbf{L}}\lambda$ such that

$$\mathbf{L}w = \left(\frac{\partial w}{\partial q}, u(q) \right) + \left(\frac{\partial w}{\partial p}, u(p) \right) + \beta w,$$

$$\tilde{\mathbf{L}}\lambda = \left(\frac{\partial \lambda}{\partial x}, u(x) \right) + \frac{1}{n} \operatorname{sp} \frac{\partial u}{\partial x} + \beta \lambda,$$

where β is a constant, $\operatorname{sp} \dfrac{\partial u}{\partial x}$ is the trace of the matrix $\dfrac{\partial u_k}{\partial x_j}$, $k, j = 1, 2, \ldots, n$.

Now define M setting Mw be equal to the following determinant of the order $n+2$

$$Mw = \begin{vmatrix} 0 & 0 & \frac{\partial \mathbf{L}w}{\partial q_1} & & \frac{\partial \mathbf{L}w}{\partial q_n} \\ 0 & 0 & \frac{q_1}{|q|} & & \frac{q_n}{|q|} \\ \frac{\partial \mathbf{L}w}{\partial p_1} & \frac{p_1}{|p|} & & & \\ & & & \frac{\partial^2 \mathbf{L}w}{\partial p \partial q} & \\ \frac{\partial \mathbf{L}w}{\partial p_n} & \frac{p_n}{|p|} & & & \end{vmatrix}$$

where $\dfrac{\partial^2 \mathbf{L}w}{\partial p \partial q}$ is a matrix of the order n.

Let $\Gamma(t)$ denote the gamma function, dp, dq denote the surface area elements of the sphere $\partial\omega$ at the points p and q respectively, and let

$$\|F\| = \sup|F(q)|, \quad q \in D.$$

Theorem 4.17. *(Anikonov, 1978a) The following inequalities hold*

$$\left\|\frac{\tilde{\mathbf{L}}\lambda}{\lambda}\right\|_{C(\omega)} \geq \left\|\frac{\mathbf{L}w}{w}\right\|_{C(\partial\omega\times\partial\omega)},$$

$$\int_\omega \lambda^{n-2}(\tilde{\mathbf{L}}\lambda)^2 \mathrm{d}x \leq \left|\frac{\Gamma(n/2)}{(n-1)2\pi^{n/2}} \int_{\partial\omega}\int_{\partial\omega} Mw \mathrm{d}p\mathrm{d}q\right|.$$

From Theorem 4.17 we obtain the table

$\bar{u} = (0,0,1) = k,$	$v = F(x_1, x_2),$
$\bar{u} = [k, x],$	$v = F\left(x_1^2 + x_2^2, x_3\right),$
$\bar{u} = x,$	$v = \|x\| F\left(\frac{x}{\|x\|}\right),$
$\bar{u} = a_0 x + [k, x],$	$v = x_3 F\left(\frac{\sqrt{x_1^2+x_2^2}}{x_3}, \sqrt{x_1^2+x_2^2}\,\mathrm{e}^{-a_0 \arctan(x_2/x_1)}\right),$
$\bar{u} = \alpha k + [k, x],$	$v = F\left(\sqrt{x_1^2+x_2^2}, x_3 - \alpha \arctan\frac{x_2}{x_1}\right),$
$\bar{u} = -k\|x\|^2 + 2x(k, x),$	$v = \|x\|^2 F\left(\frac{x_1}{\|x\|^2}, \frac{x_2}{\|x\|^2}\right),$
$\bar{u} = \alpha_0 - k\|x\|^2 + 2x(k, x),$	$v = \sqrt{x_1^2+x_2^2} F\left(\frac{x_2}{x_1}, \frac{\alpha_0 x_3 + \|x\|^2}{\sqrt{x_1^2+x_2^2}}\right),$
$\bar{u} = -k\|x\|^2 + 2x(k, x) + \alpha k,$	$v = \sqrt{x_1^2+x_2^2} F\left(\frac{x_2}{x_1}, \frac{\|x\|^2 + \alpha}{\sqrt{x_1^2+x_2^2}}\right),$

where $\bar{u} = \bar{u}(x)$, $n = 3$, $v(x) = 1/\lambda(x)$, and F is an arbitrary function.

We consider a certain inverse kinematic problem connected with geometry as well.

Let $\mathbf{R}^n$ be the $n-$ dimensional Euclidean space of variables $x = (x_1, x_2, \dots, x_n)$, $n = 2, 3, \dots$, and Q be the ball $|x| \le R$, with the boundary ∂Q. Let B be an $m-$ dimensional differentiable manifold embedded in the ball Q with the local coordinates $u = (u_1, u_2, \dots, u_m)$ and the corresponding local parametrization $x = x(u)$, $|u| \le 1$, $m \le n$.

Consider the Riemann metric

$$\mathrm{d}s^2 = \lambda^2(x)(\mathrm{d}x_1^2 + \dots + \mathrm{d}x_n^2) = \lambda^2(x)\,|\mathrm{d}x|^2$$

on the ball Q and let p denote a point of the sphere ∂Q. $\mathrm{d}s^2$ induces a Riemann metric

$$\mathrm{d}\bar{s}^2 = \sum_{i=1}^{m}\sum_{j=1}^{m} \lambda_{ij}(u)\mathrm{d}u_i\mathrm{d}u_j$$

on the manifold B. Here we set

$$\mathrm{d}x_i = \sum_{j=1}^{m} \frac{\partial x_i}{\partial u_j}\mathrm{d}u_j$$

in the equality $\mathrm{d}s^2 = \lambda^2\mathrm{d}x^2$. If some integral information linked with manifold B is given on the surface of the sphere ∂Q, then the problem of investigating this manifold B with induced metric given above is of interest for geophysics. Such an information may be given as a function

$$\int_{\gamma(x(u),p)} \lambda\,|\mathrm{d}x| + a(x(u)) = g(u,p)$$

of the variables u and p, where $\gamma(x(u),p)$ is a geodesic of the metric $\mathrm{d}s^2 = \lambda^2\,|\mathrm{d}x|^2$ joining the points $x(u) \in B$ and $p \in \partial Q$. Here the functions $a = a(x)$ and $\lambda = \lambda(x)$, as well as the parametrization $x = x(u)$ are unknown a priori. The given function $g(u,p)$ may, in general, be not even continuous.

Before we formulate the results of this problem, let us give a physical interpretation of the terms above. The ball Q is the Earth, the manifold $B \subset Q$ is the set of earthquake hypocenters, $1/\lambda(x) = v(x)$ is the propagation speed of the solid waves in the Earth, $a(x)$ is the starting time of an earthquake at the hypocenter $x = x(u)$. The given function $g(u,p)$ is the "indexed" time of the arrival of the disturbance on the surface of the Earth, registered by seismometers, where u is the "index" of the earthquake, and p is the "index" of the seismometer.

Theorem 4.18. *Let the function $\lambda(x)$ be twice differentiable and suppose that through every pair of points x_1 and x_2 in the ball Q there passes a unique geodesic $\gamma(x_1, x_2)$ of the metric $\mathrm{d}s^2 = \lambda^2\,|\mathrm{d}x|^2$.*

Then the function $R(u, p_1, p_2) = g(u, p_1) - g(u, p_2)$ of the variables (u, p_1, p_2), $|u| \le 1$, $p_i \in \partial Q$, is continuously differentiable and the following formula holds:

$$\mathrm{d}\bar{u}^2 = \sum_{i=1}^{m}\sum_{j=1}^{m} \lambda_{ij}(u)\mathrm{d}u_i\mathrm{d}u_j = \frac{1}{4}\left[\max_{p_1,p_2}\sum_{j=1}^{m}\frac{\partial R}{\partial u_j}\right]^2.$$

From Theorem 4.18 it is clear that the intrinsic geometry of the Riemannian manifold B is determined by the function $g(u,p)$. This leads to corollaries of an interest for applications.

COROLLARY 4.1. *Let the dimension of the space $\mathbf{R}^n$ be three, and B be a closed convex level surface of a function $\lambda = \lambda(x)$; that is, $B = \{\lambda(x) = c\}$. Then the surface B is uniquely determined by the function $g(u,p)$ to within the similarity and position in the ball Q.*

COROLLARY 4.2. *Let $m = n$ and let B be a simply-connected region with a smooth boundary; then both the function $\lambda = \lambda(x)$, $x \in B$, and the region B are uniquely determined by the function $g(u,p)$ to within the conformal transformations of the region B.*

Given additional hypotheses on B and the function $a(x)$ we may somewhat strengthen Theorem 4.18, namely

Theorem 4.19. *Let B be a simply-connected n-dimensional region with a smooth boundary and let the function $a(x) = 0$, $x \in B$. Then the elements of the matrix $\mathbf{c}_{ij}(u)$ inverse to $\lambda_{ij}(u)$ satisfy the relationship*

$$\sum_{i=1}^{n}\sum_{j=1}^{n} \frac{\partial g}{\partial u_i}\frac{\partial g}{\partial u_j}\mathbf{c}_{ij}(u) = 1,$$

for all $p \in \partial Q$ and, therefore, are uniquely determined by the set $g^k(u) = g(u,p_k)$ of values of the function $g(u,p)$ at the points $p = p_k \in \partial Q$ where the determinant $\left(\frac{\partial g^k}{\partial u_i}\right)\left(\frac{\partial g^k}{\partial u_j}\right)$ is nonzero, $k = 1,2,\ldots,n(n+1)$, $\quad i,j = 1,2,\ldots,n$.

Here the function $\lambda(x)$, $x \in B$ and the region B are unique to within the conformal maps.

4.4 THE STRUCTURE OF RIEMANN SPACES AND PROBLEMS OF THE INTEGRAL GEOMETRY

The aim of this section is to formulate and prove the necessary and sufficient condition for the resolvability of Riemann spaces, and the existence of certain structures in these spaces, using the integral information associated with the geodesics. We give a detailed analysis of the results reported in section 4.2.6.

The proofs are based on the differential identities and theorems of uniqueness and stability of solutions to problems in integral geometry. The results can be used, in particular, by determining the Riemann metric by the lengths of geodesics which connect the boundary points of a smooth Riemann manifold. This problem is sometimes called the problem of nonlinear tomography. Its solution can be appreciably simpler if the Riemann space has an additional structure, e. g., a primary integral of geodesics. We shall establish the criteria for the existence of a primary integral which is a homogeneous polynomial with respect to the vector tangent of the geodesic. The existence of this integral is equivalent to the conservation of some symmetric covariant tensor field by the flow of geodesics, or (which is the same) to the identity $du = 0$, where d denotes the symmetric part of the covariant derivative. Besides, we shall consider the symmetric tensor fields which satisfy the equation $du = \sigma(\mathbf{g} \otimes \mathbf{g} \otimes \ldots \otimes \mathbf{g})$ where σ denotes symmetrization, $\mathbf{g}$ is the metric tensor, and $\otimes$ denotes the tensor product.

The criteria for the existence of these fields on manifolds with a convex boundary of nonpositive curvature are given in terms of the lengths of the geodesics connecting the boundary points. In conclusion we shall formulate a criterion for the zero Gaussian curvature of a two-dimensional Riemann metric; it is based on the so-called turning number of geodesic which connect the boundary points of a circle.

4.4.1. Designations, preliminary considerations and results.

We assume that M is a smooth $n-$ dimensional Riemann manifold and T is a tangent fibre bundle for M. Let us denote the points of T by (x,ξ), where $x \in M$ and $\xi \in T_x$ (T_x is a tangent space at the point x). We use coordinate systems

$$(\pi^{-1}U, x^1, \ldots, x^n, \xi^1, \ldots, \xi^n)$$

on T, where $U \subset M$ is the domain of definition of the coordinates $(U, x^1, \ldots, x^n)$ on M, π is a projection $T \to M$, and ξ^i are the coordinates of the vector $\xi \in T_x$ with respect to the basis $\partial/\partial x^i$, $i = 1, \ldots, n$.

Let $|\xi|$ denote the length of a vector $\xi \in T_x$, $|\xi| = \sqrt{g_{ij}(x)\xi^i\xi^j}$, where $g_{ij}(x)$ are covariant components of the metric tensor. We use the conventional rule which implies summation for recurring subscripts and superscripts. Let us consider the submanifold Ω of unit vectors in T:

$$\Omega = \{(x,\xi) \in T : \xi \in \Omega_x\},$$

where Ω_x is a unit sphere at the point x, and use on Ω the same coordinates as on T, bearing in mind that $|\xi| = 1$.

We shall introduce some objects of tensor analysis on the tangent fibre bundle for the manifold M (see (Pestov and Sharafutdinov, 1988)).

A tensor of type (r,s) on the manifold T is called a semi-basic tensor if it can be represented as

$$u = u^{i_1 \ldots i_r}_{j_1 \ldots j_s} \frac{\partial}{\partial \xi^{i_1}} \otimes \ldots \otimes \frac{\partial}{\partial \xi^{i_r}} \otimes \mathrm{d}x^{j_1} \otimes \ldots \otimes \mathrm{d}x^{j_s} \tag{4.6}$$

in each coordinate system $(\pi^{-1}U, x^1, \ldots, x^n, \xi^1, \ldots, \xi^n)$ on T. Here $u^{i_1 \ldots i_r}_{j_1 \ldots j_s} \in C^\infty(\pi^{-1}U)$. The coefficients in (4.6) are supposed to be real functions. By B^r_s we denote the fibre bundle of $(r,s)-$ tensors on T, and $C^\infty(T, B^r_s)$ denotes the set of sections in this fibre bundle, i. e. the set of smooth semi-basic $(r,s)-$ tensor fields on T. If N is a submanifold in T, we denote the restriction of $C^\infty(T, B^r_s)$ on N by $C^\infty(N, B^r_s)$. In the scalar case $r = s = 0$, and we use the designation $C^\infty(N)$. We can extend the conventional algebraic tensor operations, such as addition, tensor multiplication, or convolution, to semi-basic tensors. The Riemann metric determines the operation of changing from subscripts to superscripts and *vice versa*, and allows us to introduce a scalar product

$$(u,v) = u_{i_1 \ldots i_m} v^{i_1 \ldots i_m}, \quad (u,u) = |u|^2,$$

for tensors of equal valency $m = r + s$.

The differential operators of the first order $\overset{h}{\nabla}$ and $\overset{v}{\nabla}$: $C^\infty(T, B^r_s) \to C^\infty(T, B^r_{s+1})$ can be defined in the coordinate form as

$$\overset{v}{\nabla}_k u^{(r)}_{(s)} = \frac{\partial}{\partial \xi^k} u^{(r)}_{(s)}, \quad \overset{h}{\nabla}_k u^{(r)}_{(s)} = \nabla_k u^{(r)}_{(s)} - \Gamma^p_{kl} \xi^l \frac{\partial}{\partial \xi^p} u^{(r)}_{(s)},$$

where Γ^p_{kl} are the Christoffel symbols of the metric tensor $\mathbf{g}_{ij}$; $\nabla_k u^{(r)}_{(s)}$ can be found by the rules of covariant differentiation for the ordinary $(r,s)-$ tensor fields on M. The operators $\overset{h}{\nabla}$ and $\overset{v}{\nabla}$ are called the vertical and the horizontal covariant derivatives (see (Pestov and Sharafutdinov, 1988)), respectively. Hereafter, we shall use these terms. The following formulas for these operators can be directly verified:

$$\overset{v}{\nabla}_k \overset{v}{\nabla}_l - \overset{v}{\nabla}_l \overset{v}{\nabla}_k = \overset{v}{\nabla}_k \overset{h}{\nabla}_l - \overset{h}{\nabla}_l \overset{v}{\nabla}_k = 0,$$

$$\left(\overset{v}{\nabla}_k \overset{h}{\nabla}_l - \overset{h}{\nabla}_l \overset{v}{\nabla}_k\right) u = -\mathbf{R}^p_{qkl}\xi^q \overset{v}{\nabla}_p u, \quad u \in C^\infty(T), \tag{4.7}$$

$$\overset{v}{\nabla}_k \mathbf{g}_{ij} = \overset{h}{\nabla}_k \mathbf{g}_{ij} = 0, \quad \overset{v}{\nabla}_k \delta^i_j = \overset{h}{\nabla}_k \delta^i_j = 0, \quad \overset{h}{\nabla}_j \xi^i = 0, \quad \overset{v}{\nabla}_j \xi^i = \delta^i_j,$$

where $(\mathbf{R}^i_{jkl})$ is a curvature tensor. We define contravariant derivatives

$$\overset{h}{\nabla}{}^i = g^{ij} \overset{h}{\nabla}_j, \quad \overset{v}{\nabla}{}^i = g_{ij} \overset{v}{\nabla}_j,$$

where g^{ij} are the contravariant components of the metric tensor.

The operator $\overset{h}{\nabla}$ can also be considered on the fields which belong to $C^\infty(\Omega, B^r_s)$, because $\overset{h}{\nabla}_i |\xi| = 0$ and, consequently, $\overset{h}{\nabla}_i$, $i = 1, \ldots, n$, are vectors tangent to Ω for all $(x, \xi) \in \Omega$. Besides, there are tangent vectors P_i on Ω, which are defined by the decomposition

$$\overset{v}{\nabla}_i = \frac{1}{|\xi|^2}\xi_i R + P_i, \quad R = \xi^i \overset{v}{\nabla}_j, \quad P_i = \frac{1}{|\xi|^2}\xi^j(\xi_j \overset{v}{\nabla}_i - \xi_i \overset{v}{\nabla}_j), \tag{4.8}$$

where $\xi_i = g_{ij}\xi^j$, and hence the operator

$$\mathbf{P} : C^\infty(\Omega, B^r_s) \to C^\infty(\Omega, B^r_{s+1}), \quad (\mathbf{P}u)^r_{s+1} = \left(\mathbf{P}_i u^{(r)}_{(s)}\right)$$

is defined correctly. Let $\mathbf{P}^i = g_{ij}\mathbf{P}_j$.

The ordinary tensor fields on M can be identified with semi-basic fields whose components are independent of ξ. We shall consider symmetric tensor fields on M, i. e. $(0, m)-$ tensor fields whose covariant components are symmetric about subscript transpositions.

Let S_m be a fibre bundle of symmetric tensors on M with valency m, and $C^\infty(M, S_m)$ be a set of smooth symmetric tensor fields on M. The restriction of $C^\infty(M, S_m)$ on ∂M is denoted by $C^\infty(\partial M, S_m)$.

By $d : C^\infty(M, S_m) \to C^\infty(M, S_{m+1})$ we denote the symmetric covariant derivative (see (Sharafutdinov, 1984)), $d = \sigma\nabla$, where σ denotes symmetrization, and ∇ is the operator of covariant derivative. We define the divergence $\delta : C^\infty(M, S_m) \to C^\infty(M, S_{m-1})$, $m \geq 1$, as

$$(\delta u)_{i_1 \ldots i_{m-1}} = g^{kl}\nabla_k u_{i_1 \ldots i_{m-1}}, \qquad u \in C^\infty(M, S_m).$$

The operator δd is a natural generalization of the Laplacian for symmetric tensor fields, in particular, the Dirichlet problem

$$\delta d u = f, \quad u|_{\partial M} = u_0$$

has a unique solution for any $u_0 \in C^\infty(\partial M, S_m)$ and $f \in C^\infty(M, S_m)$ (see (Pestov and Sharafutdinov, 1988)).

For symmetric tensors **u** and **v** we define symmetric product $\mathbf{uv} = \sigma(\mathbf{u} \otimes \mathbf{v})$, and denote the symmetric power of a tensor **u** by

$$\mathbf{u}^m = \sigma(\overbrace{\mathbf{u} \otimes \mathbf{u} \otimes \ldots \otimes \mathbf{u}}^{m}).$$

Let $u \in C^\infty(M, S_m)$, $m \geq 1$. Then $< u, \xi >$ will denote a function on T (or on Ω) which is defined in coordinate notation as

$$< u, \xi >= u_{i_1 \ldots i_m}(x)\xi^{i_1} \ldots \xi^{i_m}, \quad \xi \in T_x, \quad \xi \in \Omega_x.$$

Let $H = \xi^j_j \overset{h}{\nabla}$ be the geodesic flow of a manifold M (see (Godbillon, 1969)), and let δ^h and δ^v be horizontal and vertical divergence, respectively,

$$\delta^h u = \overset{h}{\nabla}_i u^i, \quad \delta^v u = \overset{v}{\nabla}_i u^i.$$

Of crucial importance for the proof is the following identity.

Lemma 4.1. *Let $w \in C^\infty(T)$. Then the identity*

$$\left|\overset{h}{\nabla} w\right|^2 - \mathbf{R}_{ijkl}\xi^i\xi^k \overset{v}{\nabla}^j w \overset{v}{\nabla}^l w + \delta^v(\overset{h}{\nabla} w H w) = \delta^h(V + 2w \overset{v}{\nabla} H w) - 2w\delta^h(\overset{v}{\nabla} H w) \quad (4.9)$$

is true for a semi-basic vector field V which is defined by

$$V^i = \overset{h}{\nabla}_j w(\xi^j \overset{v}{\nabla}^j w - \xi^i \overset{v}{\nabla}^j w)$$

PROOF. By definition,

$$\xi^i \overset{h}{\nabla}_i w = H w.$$

As a result of the action of the operator $\overset{h}{\nabla}^j w \overset{v}{\nabla}_j$, we have

$$\left|\overset{h}{\nabla} w\right|^2 - \xi^i \overset{h}{\nabla}^j w \overset{v}{\nabla}_j \overset{h}{\nabla}_i w = \overset{h}{\nabla}^j w \overset{v}{\nabla}_j H w = \delta^h(w \overset{v}{\nabla} H w) - w\delta^h(\overset{v}{\nabla} H w). \quad (4.10)$$

Let us rearrange the second term on the left-hand side of this equation. According to (4.7), we obtain.

$$2\xi^i \overset{h}{\nabla}^j w \overset{v}{\nabla}_j \overset{h}{\nabla}_i w = 2\xi^i \overset{h}{\nabla}^j w \overset{h}{\nabla}_i \overset{v}{\nabla}_j w = \overset{h}{\nabla}_i(\xi^i \overset{h}{\nabla}^j w \overset{v}{\nabla}^j w) + \overset{v}{\nabla}_j(\xi^i \overset{h}{\nabla}^j w \overset{h}{\nabla}_i w)$$
$$-\xi^i \overset{v}{\nabla}^j w(\overset{h}{\nabla}_i \overset{h}{\nabla}_j w - \overset{h}{\nabla}_j \overset{h}{\nabla}_i)w - \xi^i \overset{v}{\nabla}^j w \overset{h}{\nabla}_j \overset{h}{\nabla}_i)w - \left|\overset{h}{\nabla} w\right|^2 - \xi^i \overset{h}{\nabla}_i w \overset{h}{\nabla}_j \overset{v}{\nabla}^j w.$$

Here the first, fourth and sixth terms form $\delta^h V$. Taking commutation formulas (4.7) into account we find

$$2\xi^i \overset{h}{\nabla}^j w \overset{v}{\nabla}_j \overset{h}{\nabla}_i w = \delta^h V + \delta^v(\overset{h}{\nabla} w \cdot H w) + \mathbf{R}_{ijkl}\xi^i\xi^k \overset{v}{\nabla}^j w \overset{v}{\nabla}^l w - \left|\overset{h}{\nabla} w\right|^2.$$

Substituting this expression in (4.10) we obtain (4.9).

Let $\lambda \in \mathbf{R}^1$. We define a differential operator

$$P_\lambda = \overset{h}{\nabla}^i \mathbf{P}_i + \lambda H,$$

and an operator of the homogeneous extension of power λ

$$R_\lambda w = |\xi|^\lambda w\left(x, \frac{\xi}{|\xi|}\right), \quad \xi \in T_x, \quad \xi \neq 0, \quad w \in C^\infty(\Omega).$$

We obtain a similar identity for functions in $C^\infty(\Omega)$, as a corollary of Lemma 4.1.

Lemma 4.2. *For each smooth function $w(x,\xi)$ on Ω*

$$\begin{aligned} \left|\overset{h}{\nabla} w\right|^2 - \mathbf{R}_{ijkl}\xi^i\xi^k\mathbf{P}^j w\mathbf{P}^l w + \mathbf{P}_i(Hw \overset{h}{\nabla}{}^i w) + (2\lambda+1)(Hw)^2 \\ = \delta^h\{V + 2w\,(\overset{v}{\nabla} R_\lambda Hw)\Big|_\Omega\} - 2wP_\lambda Hw \end{aligned} \tag{4.11}$$

where

$$V^i = \overset{h}{\nabla}_j w(\xi^j\mathbf{P}^i w - \xi^i\mathbf{P}^j w).$$

PROOF. The proof consists in applying identity (4.9) to a function $R_\lambda w$ and using (4.8) (we should also take into account the properties of the curvature tensor and relations $\mathbf{P}_i R_\lambda = R_\lambda \mathbf{P}_i$, $\mathbf{R}R_\lambda = \lambda R_\lambda$).

Let M be compact oriented manifold. The differential form

$$\mathrm{d}T = g\mathrm{d}x \wedge \mathrm{d}\xi = g\mathrm{d}x^1 \wedge \ldots \wedge \mathrm{d}x^n \wedge \mathrm{d}\xi^1 \wedge \ldots \wedge \mathrm{d}\xi^n,$$

determines a volume element of the manifold T, where $g = \det(\mathbf{g}_{ij})$. A volume element $\mathrm{d}\Omega$ in Ω is defined by

$$\mathrm{d}T = \mathrm{d}\,|\xi| \wedge \mathrm{d}\Omega,$$

or

$$\mathrm{d}\Omega = \sqrt{g}\mathrm{d}x \wedge \mathrm{d}\Omega_x,$$

where $\mathrm{d}\Omega_x$ is a volume of a sphere Ω_x and $\mathrm{d}T_x = \mathrm{d}\,|\xi| \wedge \mathrm{d}\Omega_x$. In the coordinate notation we have

$$\mathrm{d}\Omega_x = \sqrt{g}(\xi^1\mathrm{d}\xi^2 \wedge \ldots \wedge \mathrm{d}\xi^n - \ldots + (-1)^{n+1}\xi^n\mathrm{d}\xi^1 \wedge \ldots \wedge \mathrm{d}\xi^{n-1}), \quad \xi \in \Omega_x.$$

Let us define on $\partial\Omega$ a differential form

$$\mathrm{d}E = \mathrm{d}S \wedge \Omega_x, \quad x \in \partial M,$$

where

$$\mathrm{d}S = \sqrt{g}(\nu^1\mathrm{d}x^2 \wedge \ldots \wedge \mathrm{d}x^n - \ldots + (-1)^{n+1}\nu^n\mathrm{d}x^1 \wedge \ldots \wedge \mathrm{d}x^{n-1})$$

is a volume element of a boundary ∂M. The form $\mathrm{d}E$ is naturally referred to as the volume form of the manifold $\mathrm{d}\Omega$, because

$$\mathrm{d}\Omega = -\mathrm{d}\rho \wedge \mathrm{d}E,$$

where $\rho(x)$ denotes the distance from the boundary ∂M. The function $\rho(x)$ is smooth in the neighbourhood of the boundary, and $\nabla\rho = -\nu$ on ∂M.

In the sequel we use the following integral relations.

Let u be a semi-basic vector field on T. The Gauss-Ostrogradskii formula,

$$\int_{\Omega} \overset{h}{\delta}\, u \mathrm{d}\Omega = \int_{\partial\Omega} (u, \nu) \mathrm{d}E, \tag{4.12}$$

holds for the horizontal divergence (see (Pestov and Sharafutdinov, 1988)). The formula

$$\int_{\Omega_x} \mathbf{P}_i u^i \mathrm{d}\Omega_x = (n-1) \int_{\Omega_x} (\xi, u) \mathrm{d}\Omega_x \tag{4.13}$$

follows from the equality

$$\int_{\Omega_x} v \xi_i \mathrm{d}\Omega_x = \int_{|\xi| \leq 1} v_{\xi^i} \mathrm{d}T_x, \quad v \in C^\infty(T_x). \tag{4.14}$$

Indeed, we have

$$\int_{\Omega_x} \xi^k \left(u_{\xi^i}\xi_k - u_{\xi^k}\xi_i\right) \mathrm{d}\Omega_x = \int_{|\xi| \leq 1} \left[\left(\xi^k u_{\xi^i}\right)_{\xi}^{k} - \left(\xi^k u_{\xi^k}\right)_{\xi}^{i}\right] \mathrm{d}T_x$$

$$= (n-1) \int_{|\xi| \leq 1} u_{\xi}^{i} \mathrm{d}T_x = (n-1) \int_{\Omega_x} \xi_i u \mathrm{d}\Omega_x,$$

and hence we obtain (4.13) for the vector field.

The operator of integration over the sphere Ω_x transforms the semi-basic tensor fields on Ω to ordinary tensor fields on M. In particular, let us consider the symmetric tensor field on M:

$$\int_{\Omega_x} \xi^{i_1} \ldots \xi^{i_{2m}} \mathrm{d}\Omega_x.$$

We shall know that the equation

$$\int_{\Omega_x} \xi^{i_1} \ldots \xi^{i_{2m}} \mathrm{d}\Omega_x = c_m (\mathbf{g}^m)^{i_1 \ldots i_{2m}} \tag{4.15}$$

is true, where $\mathbf{g}^m$ the $m-$ th symmetric power of the metric tensor $\mathbf{g}^{ij}$,

$$\mathbf{g}^m = \sigma(\overbrace{\mathbf{g} \otimes \mathbf{g} \otimes \ldots \otimes \mathbf{g}}^{m})$$

and

$$c_m = \frac{1 \cdot 3 \cdot \ldots \cdot (2m-1)}{n(n+2)\ldots(n+2m-2)} \sigma_n,$$

where σ_n is the the volume of unit sphere in $\mathbf{R}^n$. We prove (4.15) by induction.

Let $m = 1$. We easily see (e.g. in Riemann coordinates in the neighbourhood of a point x) that

$$\int_{\Omega_x} \xi^i \xi^j \mathrm{d}\Omega_x = c\mathbf{g}^{ij},$$

where the constant c can be defined from convolution

$$\int_{\Omega_x} d\Omega_x = \sigma_n = cn, \quad c = c_1 = \frac{\sigma_n}{n}.$$

Suppose that (4.15) is established for some m. Then, according to (4.14),

$$\int_{\Omega_x} \xi^{i_1} \dots \xi^{i_{2m+1}} \xi_{i_{2m+2}} d\Omega_x$$

$$= \int_{|\xi|\leq 1} \left[\delta^{i_1}_{i_{2m+2}} \xi^{i_2} \dots \xi^{i_{2m+1}} + \delta^{i_2}_{i_{2m+2}} \xi^{i_1} \xi^{i_3} \dots \xi^{i_{2m+1}} + \dots + \delta^{i_{2m+1}}_{i_{2m+2}} \xi^{i_1} \dots \xi^{i_{2m+1}}\right] dT_x.$$

Taking into account the homogeneity of the terms in the right-hand side, we rearrange the latter equation:

$$\int_{\Omega_x} \xi^{i_1} \dots \xi^{i_{2m+1}} \xi_{i_{2m+2}} d\Omega_x = \frac{1}{n+2m} \int_{\Omega_x} \left[\delta^{i_1}_{i_{2m+2}} \xi^{i_2} \dots \xi^{i_{2m+1}} + \delta^{i_{2m+1}}_{i_{2m+2}} \xi^{i_1} \dots \xi^{i_{2m+1}}\right] d\Omega_x.$$

Changing from subscript to superscript and using the induction assumption, we get

$$\int_{\Omega_x} \xi^{i_1} \dots \xi^{i_{2m+2}} d\Omega_x = \frac{c_m}{n+2m} \left[\mathbf{g}^{i_1 i_{2m+2}} (\mathbf{g}^m)^{i_2 \dots i_{2m+1}} + \mathbf{g}^{i_{2m+1} i_{2m+2}} (\mathbf{g}^m)^{i_1 \dots i_{2m+1}}\right]$$

$$= c_{m+1} (\mathbf{g}^{m+1})^{i_1 \dots i_{2m+2}}.$$

Suppose that M is a compact dispersive manifold, i. e. a compact oriented manifold which does not contain infinitely long geodesics and has a strictly convex boundary ∂M. Then, we can define on Ω a function $t(x, \xi)$ as the length of a geodesic $\gamma(x, \xi, t)$ which emerges from a point $x \in M$ in the direction of a vector $\xi \in \Omega_x$. Here Ω is a compact manifold with boundary $\partial\Omega = \partial_-\Omega \cap \partial_+\Omega$, where

$$\partial_-\Omega = \{(x, \xi) \in \Omega : x \in \partial M, \quad (\xi, \nu(x)) \leq 0\},$$

$$\partial_+\Omega = \{(x, \xi) \in \Omega : x \in \partial M, \quad (\xi, \nu(x)) \geq 0\}.$$

and $\nu(x)$ is the unit outward normal of ∂M at the point x. We also designate

$$\partial_0\Omega = \{(x, \xi) \in \Omega : (\xi, \nu(x)) = 0\}.$$

Obviously, $\partial_0\Omega = \partial(\partial_-\Omega) = \partial(\partial_+\Omega)$. By definition, $t(x, \xi) = 0$ for $(x, \xi) \in \partial_+\Omega$. The restriction of the function $t(x, \xi)$ on $\partial_-\Omega$ will be denoted by $t_0(x, \xi)$. We define a mapping $\psi : \partial_-\Omega \to \partial_-\Omega$:

$$\psi(x, \xi) = \{\gamma\left[x, \xi, t_0(x, \xi)\right], -\dot{\gamma}\left[x, \xi, t_0(x, \xi)\right]\}.$$

As above, let $\rho(x)$ be the distance from the boundary ∂M. In the region where $\rho(x)$ is smooth we consider the following decomposition similar to (4.13):

$$\nabla\rho = (\nabla\rho, \overset{h}{\nabla})\nabla\rho + Q,$$

where

$$Q_i = \nabla^k\rho(\nabla_k\rho \overset{h}{\nabla}_i - \nabla_i\rho \overset{h}{\nabla}_k).$$

The main properties of the function $t(x, \xi)$ are stated in the following lemma.

Lemma 4.3. *Let M be a compact dispersive manifold. Then*

1. $t(x,\xi) \in C^\infty(\Omega\backslash\partial_0\Omega)$,

2. $Ht = -1$, (4.16)

3. $t_0 \in C^\infty(\partial_-\Omega)$,

4. *ψ is a diffeomorphism $\partial_-\Omega$ to $\partial_-\Omega$ and $\psi \circ \psi = id$, where id denotes the identity mapping,*

5. *the derivatives $P_i t$ and $Q_i t$ are bounded in $\Omega\backslash\partial_0\Omega$.*

PROOF. The following proof of statement 1. is given in (Pestov and Sharafutdinov, 1988).

1. Let a set M_0 be defined by

$$M_0 = \{x \in M, \quad \rho(x) \geq \rho_0, \quad \rho_0 > 0\}.$$

The function $\rho(x)$ is smooth in a certain neighbourhood of the boundary. Let us assume that ρ_0 is so small that a set $M\backslash M_0$ can be considered as this neighbourhood; it is evidently a smooth manifold. Hereafter we consider $\rho(x)$ only on $M\backslash M_0$. The function $t(x,\xi)$ is a solution of the equation

$$\rho(\gamma(x,\xi,t(x,\xi))) = 0.$$

Suppose that $\cos\varphi(x,\xi)$ corresponds to the angle between a geodesic and ∂M, i. e.

$$\cos\varphi(x,\xi) = -\frac{\partial\rho(\gamma(x,\xi,t))}{\partial t},$$

where $t = t(x,\xi)$. Since $\gamma(x,\xi,t)$ is a smooth curve which smoothly depends on the initial data, then, according to the implicit function theorem, $t(x,\xi)$ is a smooth function for all (x,ξ), where $\cos\varphi(x,\xi) \neq 0$. This inequality holds for all $(x,\xi) \in \Omega\backslash\partial_0\Omega$, provided that the boundary is strictly convex.

2. Let (x,ξ) be an interior point in the manifold Ω. We define the operator $\mathbf{h}_t$ of translation along the geodesic flow:

$$\mathbf{h}_t(x,\xi) = (\gamma(x,\xi,t), \quad \dot\gamma(x,\xi,t)), \quad -t(x,-\xi) \leq t \leq t(x,\xi).$$

Direct calculation based on the equation for geodesic, $\ddot\gamma^i + \Gamma^i_{jk}\dot\gamma^j\dot\gamma^k$, leads to

$$\frac{\mathrm{d}}{\mathrm{d}t} - (u \circ \mathbf{h}_t) = (Hu) \circ \mathbf{h}_t, \quad u \in C^\infty(\Omega). \tag{4.17}$$

Applying (4.17) to $t(x,\xi)$ and assuming $t = 0$ we obtain (4.16) (we take into account that $t(\mathbf{h}_t(x,\xi)) = t(x,\xi) - t$ and $\mathbf{h}_0 = id$).

3. Let t_0^- denote that part of the function $t|_{\partial\Omega}$ which is an odd function in ξ. It is evident that $t_0^- = t_0/2$ on $\partial_-\Omega$. We are to show that $t_0^- \in C^\infty(\partial\Omega)$ and thus prove statement 3.

Let us consider the function

$$\mu(x,\xi,t) = \tilde\rho(\gamma(x,\xi,t)), \quad (x,\xi) \in \partial\Omega, \quad -t(x,-\xi) \leq t \leq t(x,\xi),$$

where $\tilde{\rho}(x)$ is a smooth extension of $\rho(x)$ to the entire manifold M ($\rho(x)$ is defined only on $M\backslash M_0$). Integrating the identity $(t\dot{\mu})' = \dot{\mu} + t\,\ddot{\mu}$ from $-t(x,-\xi)$ to $t(x,\xi)$ we have

$$2t_0^-(x,\xi)\dot{\mu}(x,\xi,2t_0^-(x,\xi)) = \int\limits_0^{t_0^-(x,\xi)} t\ddot{\mu}(x,\xi,t)dt.$$

Changing the integration variable and dividing by $2t_0^-$ we obtain

$$F(x,\xi,2t_0^-) \equiv \dot{\mu}(x,\xi,2t_0^-) - 2t_0^- \int\limits_0^1 s\ddot{\mu}(x,\xi,2st_0^-)ds = 0.$$

It follows from 1. that the function $t|_{\partial Q}$ is smooth on $\partial Q\backslash\partial_0 Q$. Therefore, it is enough to establish that t_0^- is a smooth function in the neighbourhood of any point $(x_0,\xi_0) \in \partial_0\Omega$. The function $F(x,\xi,t)$ is defined and smooth for $(x,\xi) \in \Omega$, $-t(x,-\xi) \le t \le t(x,\xi)$ and satisfies the conditions

$$F(x_0,\xi_0,0) = -(\nu(x_0),\xi_0) = 0,$$

$$F_t(x_0,\xi_0,0) = \frac{1}{2}\,\ddot{\mu}(x_0,\xi_0,0) = \frac{1}{2}\,\xi_0^i\xi_0^j\nabla_i\nabla_j\rho(x_0) < 0.$$

The latter inequality implies that the boundary ∂M is strictly convex. According to the implicit function theorem, the smoothness of F implies that t_0^- is also smooth in the neighbourhood of (x_0,ξ_0), and consequently t_0 is smooth on $\partial\Omega$.

4. The fact that ψ is a bijection follows from the uniqueness of the solution to the Cauchy problem for geodesics, and 3. implies that the mapping ψ is smooth. Equation $\psi\circ\psi = id$ is evident.

5. The ratio $t_0(x,\xi)/\cos\varphi(x,\xi)$ is bounded on $\partial_-\Omega\backslash\partial_0\Omega$ (Pestov and Sharafutdinov, 1988). This implies that the function $t(x,\xi)/\cos\varphi(x,\xi)$ is bounded in $\overset{\circ}{\Omega}$, where $\overset{\circ}{\Omega}$ is the interior of the manifold Ω.

Let $(x,\xi) \in \overset{\circ}{\Omega}$. Consider a point $(x_1,\xi_1) \in \partial_-\Omega$ with:

$$x_1 = \gamma(\dot{x},\xi,-t(x,-\xi)), \quad \xi_1 = -\dot{\gamma}(x,\xi,-t(x,-\xi)).$$

We have

$$\frac{t(x_1,\xi_1)}{\cos\varphi(x_1,\xi_1)} < c,$$

but $\cos\varphi(x_1,\xi_1) = \cos\varphi(x,\xi)$ and $t(x,\xi) < t(x_1,\xi_1)$, therefore

$$\frac{t(x,\xi)}{\cos\varphi(x,\xi)} < c.$$

Similar to 3. we introduce a function

$$\mu(x,\xi,t) = \tilde{\rho}(\gamma(x,\xi,t)).$$

Here we consider this function on the set $(x,\xi) \in \overset{\circ}{\Omega}$, $0 \le t \le t(x,\xi)$. Integrating the identity

$$(t\dot{\mu})' = \dot{\mu} + t\ddot{\mu}$$

from zero to $t(x,\xi)$ we obtain

$$-t(x,\xi)\cos\varphi(x,\xi) = -\tilde{\rho}(x) + \int\limits_0^{t(x,\xi)} t\tilde{\mu}(x,\xi,t)\mathrm{d}t. \tag{4.18}$$

Since $t(x,\xi) \in C^\infty(\overset{\circ}{\Omega})$, it is enough to verify statement 5. for $x \in \overset{\circ}{M} \setminus M_0$, where $\overset{\circ}{M}$ is the interior of M. For these x we can replace $\tilde{\rho}(x)$ in (4.18) by $\rho(x)$. Let us apply the operator $\mathbf{P}_i$ (or $\mathbf{Q}_i$) $i = 1,\dots,n$ to (4.18) (note that $\mathbf{P}_i\rho = \mathbf{Q}_i\rho = 0$):

$$-\cos\varphi(x,\xi)\mathbf{P}_i t(x,\xi) + t(x,\xi)\left[(\mathbf{P}_i\mu)(x,\xi,t(x,\xi)) + \tilde{\mu}(x,\xi,t(x,\xi))\mathbf{P}_i t(x,\xi)\right]$$

$$= t(x,\xi)\tilde{\mu}(x,\xi,t(x,\xi))\mathbf{P}_i t(x,\xi) + \int\limits_0^{t(x,\xi)} t\mathbf{P}_i\tilde{\mu}(x,\xi,t)\mathrm{d}t.$$

This together with the boundedness of $t(x,\xi)/\cos\varphi(x,\xi)$ imply that the derivatives $\mathbf{P}_i t(x,\xi)$ and $\mathbf{Q}_i t(x,\xi)$, $i = 1,\dots,n$, are also bounded on $\overset{\circ}{\Omega}$ (in any coordinate system).

4.4.2. Formulation and proof of the main results

We assume that M is a compact dispersive manifold and

$$u \in C^\infty(M, S_m), \quad m \geq 1.$$

Calculations lead to

$$H < u,\xi > = < \mathrm{d}u,\xi >, \tag{4.19}$$

or, taking into account (4.16)

$$\frac{\mathrm{d}}{\mathrm{d}t} < u(\gamma),\dot{\gamma} > = < \mathrm{d}u(\gamma),\dot{\gamma} >, \tag{4.20}$$

where $\gamma(x,\xi,t)$. For $\mathrm{d}u = 0$ this implies that $< u(\gamma),\dot{\gamma} >$ is a primary integral for the equation of geodesics. In particular, for $(x,\xi) \in \partial_-\Omega$, $t = t(x,\xi)$ we obtain

$$< u_0,\xi > = (-1)^m < u_0,\xi > \circ\psi,$$

where $u_0 = u|_{\partial M}$. Now the question arises whether there exists a smooth symmetric field u of valency $m \geq 1$ on M such that $\mathrm{d}u = 0$, provided

$$< u_0,\xi > = (-1)^m < u_0,\xi > \circ\psi$$

on $\partial_-\Omega$ for field $u_0 \in C^\infty$. We answer this question, as well as a similar question for the equality $\mathrm{d}u = g^k$, positively. This is one of the main results of this section.

Theorem 4.20. *Let M be a compact dispersive manifold with nonpositive curvature, i. e.*

$$\mathbf{R}_{ijkl}\xi^i\eta^j\xi^k\eta^l \leq 0$$

for all $x \in M\xi$, $\eta \in T_x$ and let $u \in C^\infty(M, S_m)$, $m \geq 1$. Then the following statements are equivalent:

1. $\delta \mathrm{d}u = 0, \quad <u_0, \xi> = (-1)^m <u_0, \xi> \circ \psi, \quad u_0 = u|_{\partial M},$

2. $\mathrm{d}u = 0.$

Theorem 4.21. *Let M be a compact dispersive manifold with nonpositive curvature, and $u \in C^\infty(M, S_{2k-1})$, $k \geq 1$. Then the following statements are equivalent:*

1. $\delta \mathrm{d}u = 0, \quad t_0 + <u_0, \xi> + <u_0, \xi> \circ \psi = 0, \quad u_0 = u|_{\partial M},$

2. $\mathrm{d}u = g^k.$

Theorems 4.1 and 4.2 are corollaries of the following result, which generalizes the main result in (Pestov and Sharafutdinov, 1988).

Theorem 4.22. *Let M be a dispersive manifold with nonpositive curvature, and*

$$w(x,\xi) = \int_0^{t(x,\xi)} f(\mathbf{h}_t(x,\xi))\mathrm{d}t,$$

where $f \in C^\infty(\Omega)$ and $P_\lambda f = 0$ for some $\lambda > -\dfrac{n+1}{2}$. Then

$$w_0 = w|_{\partial_-\Omega} \in C^\infty(\partial_-\Omega)$$

and

$$\|f\|^2_{L_2} = \frac{1}{2\lambda + n + 1} \int_{\partial_-\Omega} \left[l(\mathbf{P}w_0, \mathbf{Q}w_0) + 2w_0 \left(\nu, \overset{v}{\nabla} R_\lambda f\right)\Big|_\Omega \right] \mathrm{d}E, \tag{4.21}$$

where

$$l(\mathbf{P}w_0, \mathbf{Q}w_0) = (\xi^i \mathbf{P}^j w_0 - \xi^j \mathbf{P}^i w_0)\nu_j \mathbf{Q}_i w_0.$$

PROOF. The function w_0 is smooth because t_0 is smooth (Lemma 4.1). We shall show that w satisfies

$$Hw = -f \tag{4.22}$$

in $\overset{\circ}{\Omega}$. Let $(x,\xi) \in \overset{\circ}{\Omega}$. Then

$$t(\mathbf{h}_\tau(x,\xi)) = t(x,\xi) - \tau, \quad w(\mathbf{h}_\tau(x,\xi)) = \int_\tau^{t(x,\xi)} f(\mathbf{h}_t(x,\xi))\mathrm{d}t$$

for all τ in the interval $0 \leq \tau \leq t(x,\xi)$. If we differentiate the latter equation with respect to τ, put $\tau = 0$ and take into account (4.17), we obtain (4.22).

By Ω_0 we denote the tangent fibre bundle for the manifold M_0 introduced in Lemma 4.3. The function $w(x,\xi)$ is smooth on Ω_0 because $t(x,\xi)$ is smooth. In particular, the derivatives $\mathbf{P}_i w$ and $\mathbf{Q}_i w$, $i = 1, \ldots n$ are bounded in $\overset{\circ}{\Omega}$ in any coordinate system. Let us apply (4.11) to the function $w|_{\Omega_0}$. With regard to the theorem conditions this leads to the inequality

$$\left|\overset{h}{\nabla} w\right|^2 + \mathbf{P}_i(Hw \cdot \overset{h}{\nabla}{}^i w) + (2\lambda+1)(Hw)^2 \leq \delta^h \left(V + 2w \left(\overset{v}{\nabla} R_\lambda Hw\right)\Big|_{\Omega_0} \right), \tag{4.23}$$

where $v^i = \overset{h}{\nabla}_j w\,(\xi^i \mathbf{P}^j w - \xi^j \mathbf{P}^i w)$. Integrating (4.23) over Ω_0 we take into account (4.12), (4.13) and the obvious inequality $\left|\overset{h}{\nabla} w\right|^2 \geq (Hw)^2$ to get

$$\|f\|^2_{L_2} = \frac{1}{2\lambda + n + 1} \int\limits_{\partial\Omega_0} \left[l(\mathbf{P}w, \mathbf{Q}w) + 2w\,(\nu_0, \overset{v}{\nabla} R_\lambda f)\Big|_{\Omega_0} \right] \mathrm{d}E_0,$$

where ν_0 is the outward normal to ∂M_0, and $\mathrm{d}E_0$ is a volume element of the manifold $\partial\Omega_0$. Let us show that (4.21) can be obtained from this inequality for $\rho_0 \to 0$.

Let

$$\partial_\varepsilon \Omega_0 = \{(x,\xi) \in \partial\Omega : |(\xi, \nu_0)| \leq \varepsilon\}, \quad 0 < \varepsilon < 1.$$

We have

$$\int\limits_{\partial\Omega_0} l(\mathbf{P}w, \mathbf{Q}w)\mathrm{d}E_0 = \int\limits_{\partial_\varepsilon\Omega_0} l(\mathbf{P}w, \mathbf{Q}w)\mathrm{d}E_0 + \int\limits_{\partial\Omega_0 \backslash \partial_\varepsilon\Omega} l(\mathbf{P}w, \mathbf{Q}w)\mathrm{d}E_0.$$

Since w is smooth on $\Omega \backslash \partial_0\Omega$, the second term tends to

$$\int\limits_{\partial_-\Omega \backslash \partial_\varepsilon\Omega} l(\mathbf{P}w, \mathbf{Q}w)\mathrm{d}E,$$

when $\rho_0 \to 0$ (recall that $w|_{\partial_+\Omega} = 0$ by definition). As for the first term, its absolute value is bounded by $c\varepsilon$ with constant c independent of ρ_0, because the derivatives $\mathbf{P}_i w$ and $\mathbf{Q}_i w$, are bounded on $\overset{\circ}{\Omega}$. Hence, we see that

$$\int\limits_{\partial\Omega_0} l(\mathbf{P}w, \mathbf{Q}w)\mathrm{d}E_0 \to \int\limits_{\partial_-\Omega} l(\mathbf{P}w_0, \mathbf{Q}w_0)\mathrm{d}E$$

for $\rho_0 \to 0$. The theorem is proved.

Proof of Theorem 4.20. The fact that 2. implies 1. was proved before (Theorem 4.8). Suppose that 1. is satisfied. Let us consider a function

$$w(x,\xi) = \int\limits_0^{t(x,\xi)} f(\mathbf{h}_t(x,\xi))\mathrm{d}t$$

where $f =< \mathrm{d}u, \xi >$. The function satisfies the condition of Theorem 4.22, namely $P_{m+1} f =< \delta\mathrm{d}u, \xi >= 0$, for $\lambda = m+1$. Then, from (4.19) we have

$$w(x,\xi) =< u(\gamma(x,\xi,t(x,\xi)), \quad \dot\gamma(x,\xi,t(x,\xi)) > - < u(x), \xi >,$$

and, according to the theorem condition, $w_0 = w|_{\partial_-\Omega} = 0$. Therefore, and from (4.15) we have $\mathrm{d}u = 0$.

Proof of Theorem 4.21. Let $\mathrm{d}u = g^k$ and $u \in C^\infty(M, S_{2k-1})$, $k \geq 1$. It is easy to see that $< g^k, \xi >= 1$ if $|\xi| = 1$. Hence, according to (4.19),

$$H < u, \xi >= 1,$$

and consequently

$$H(t(x,\xi) + < u, \xi >) = 0.$$

Taking into account (4.17) we obtain

$$\frac{\mathrm{d}}{\mathrm{d}t}(w(\mathbf{h}_t(x,\xi))) = 0,$$

where $w = t(x,\xi) + (u,\xi)$. Hence

$$t_0(x,\xi) + < u_0, \xi > = - < u_0, \xi > \circ\psi$$

(here $t|_{\partial_+\Omega} = 0$). The equality $\delta \mathrm{d}u = 0$ is evident since $\nabla g = 0$.
The inverse statement can be proved similar to that of Theorem 4.20, but instead of $f(x,\xi)$ we consider the function $< \mathrm{d}u - g^k, \xi >$. Here

$$w(x,\xi) = \int_0^{t(x,\xi)} f(\mathbf{h}_t(x,\xi))\mathrm{d}t = < u(\gamma), \quad \dot\gamma > - < u(x), \quad \xi > -t(x,\xi).$$

Further we need to verify that the conditions of Theorem 4.22 are satisfied. As a result, we obtain $\mathrm{d}u = g^k$.

The problem of describing Riemann spaces which admit symmetric tensor fields satisfying $\mathrm{d}u = 0$ or $\mathrm{d}u = g^m$ remains unsolved, except for the cases $m = 1$ and $m = 2$ (Favard, 1957).

4.4.3. Two-dimensional case

In a two-dimensional case the necessary and sufficient conditions for the Gaussian curvature of regular $2D$ metric to vanish can be found in terms of the turning of the metric geodesics under weaker conditions.

We consider a positive definite two-dimensional Riemann metric of the class C^3 in the circle $x^2 + y^2 \le 1$:

$$\mathrm{d}s^2 = E(x,y)\mathrm{d}x^2 + 2F(x,y)\mathrm{d}x\mathrm{d}y + \sigma(x,y)\mathrm{d}y^2. \tag{4.24}$$

We further assume that any two points of the circle $x^2 + y^2 \le 1$ can be connected in the metric of (4.24) by a unique geodesic which depends on these points as a continuous and differentiable function.

The Gaussian curvature $k(x,y)$ of metric (4.24) is given by the following formula (Favard, 1957)

$$k = -\frac{1}{4H^2}\begin{vmatrix} E & \dfrac{\partial E}{\partial x} & \dfrac{\partial E}{\partial y} \\ F & \dfrac{\partial F}{\partial x} & \dfrac{\partial F}{\partial y} \\ \sigma & \dfrac{\partial \sigma}{\partial x} & \dfrac{\partial \sigma}{\partial y} \end{vmatrix} - \frac{1}{2H}\left[\frac{\partial}{\partial y}\frac{\left(\dfrac{\partial E}{\partial y} - \dfrac{\partial E}{\partial x}\right)}{H} - \frac{\partial}{\partial x}\frac{\left(\dfrac{\partial F}{\partial x} - \dfrac{\partial \sigma}{\partial x}\right)}{H}\right],$$

$$H = \sqrt{E\sigma - F^2}.$$

Suppose that a geodesic $\gamma(\bar a, \bar b) = (x(t), y(t))$ (in metric (4.24)) connects two points $\bar a$ and $\bar b$ in the circle $x^2 + y^2 \le 1$. Then the number

$$P(a,b) = \int_{\gamma(a,b)} \mathrm{d}\varphi = \arctan \left.\frac{Hy'}{Ex' + Fy'}\right|_{(x,y)=\bar b} - \arctan \left.\frac{Hy'}{Ex' + Fy'}\right|_{(x,y)=\bar a},$$

$$\tan \varphi = \frac{Hy'(t)}{Ex'(t) - Fy'(t)}$$

is called the turning number of this geodesic. In case of the conformal Euclidean metric $E = \sigma$, $F = 0$ and

$$\tan \varphi = \frac{y'(t)}{x'(t)}.$$

From the viewpoint of inverse problems for differential equations and particularly, kinematic inverse problems (Anikonov and Pestov, 1990a), it is interesting to determine the metric in (4.24) and its structure using the integral information on geodesics connecting only the points of the circumference $x^2 + y^2 = 1$.

Let $\gamma(z_1, z_2)$ be a geodesic of metric (4.24) which connects two points z_1 and z_2 belonging to the circumference $x^2 + y^2 = 1$ and $0 \le z_1 \le 2\pi$, $0 \le z_2 \le 2\pi$. Let $P(z_1, z_2)$ denote the turning number of $\gamma(z_1, z_2)$:

$$P(z_1, z_2) = \int\limits_{\gamma(z_1,z_2)} \mathrm{d}\varphi, \quad 0 \le z_i \le 2\pi, \quad i = 1, 2.$$

Theorem 4.23. *The Gaussian curvature $k(x, y)$ of metric (4.24) is zero if and only if $P(z_1, z_2)$ can be represented as*

$$P(z_1, z_2) = f_0(z_2) - f_0(z_1),$$

where $f_0(z)$ is a continuously differentiable function of z, $0 \le z \le 2\pi$ and $f_0(0) = f_0(2\pi)$.

PROOF. Suppose that $k(x, y) = 0$ in the circle $x^2 + y^2 \le 1$. If functions $\alpha(x, y)$ and $\beta(x, y)$ are defined by

$$\alpha = -\frac{\sqrt{E}}{H}\left(\frac{\partial \sqrt{E}}{\partial y} - \frac{\partial}{\partial x}\frac{H}{\sqrt{E}}\right), \quad \beta = \frac{1}{E}\left(F\alpha + \sqrt{E}\frac{\partial}{\partial x}\frac{H}{E}\right),$$

the following equations are true (see (Favard, 1957)):

$$-\mathrm{d}\varphi = \alpha \mathrm{d}x + \beta \mathrm{d}y, \tag{4.25}$$

$$\mathrm{d}(\alpha \mathrm{d}x + \beta \mathrm{d}y) = -k(x, y) H(x, y) \mathrm{d}x \mathrm{d}y. \tag{4.26}$$

Since $k(x, y) = 0$ in the circle $x^2 + y^2 \le 1$, then from (4.26) we have

$$\mathrm{d}(\alpha \mathrm{d}x + \beta \mathrm{d}y) = 0.$$

According to the Poincaré lemma, there exists a continuously differentiable function $f(x, y)$ such that

$$\alpha \mathrm{d}x + \beta \mathrm{d}y = \mathrm{d}f, \quad x^2 + y^2 \le 1.$$

From this equation and (4.25) we obtain for γ the relation

$$-\mathrm{d}\varphi = \mathrm{d}f.$$

Integrating the latter equation along the geodesic $\gamma(z_1, z_2)$ we have

$$P(z_1, z_2) = f_0(z_2) - f_0(z_1), \tag{4.27}$$

where $f_0(z) = -f(\cos z, \sin z)$.

The inverse statement may be proved by using more sophisticated reasoning. Thus, we suppose that

$$P(z_1, z_2) = f_0(z_2) - f_0(z_1) \tag{4.28}$$

holds for all z_1, z_2, $0 \le z_i \le 2\pi$, $i = 1,2$, where $f_0(z)$ is a continuously differentiable function of z, $0 \le z \le 2\pi$. We are to prove that $k(x,y) = 0$ in the circle $x^2 + y^2 \le 1$.

Let $f(x,y)$, $x^2+y^2 \le 1$ be a continuously differentiable extension of the function $f_0(z)$, $0 \le z \le 2\pi$. Using this extension and (4.25) we can rewrite (4.28) as

$$\int_{\gamma(z_1,z_2)} \alpha \mathrm{d}x + \beta \mathrm{d}y = \int_{\gamma(z_1,z_2)} \mathrm{d}f,$$

or

$$\int_{\gamma(z_1,z_2)} \left(\alpha - \frac{\partial f}{\partial x}\right) \mathrm{d}x + \left(\beta - \frac{\partial f}{\partial y}\right) \mathrm{d}y = 0. \tag{4.29}$$

Consequently, for any two points on the circumference $x^2 + y^2 = 1$, the integral along $\gamma(z_1, z_2)$ of the differential form

$$\omega = a\mathrm{d}x + b\mathrm{d}y, \quad a = \alpha - \frac{\partial f}{\partial x}, \quad b = \beta - \frac{\partial f}{\partial y}$$

equals zero. If we show that $\mathrm{d}\omega = 0$, the theorem will be proved. Indeed,

$$\mathrm{d}\omega = \mathrm{d}(a\mathrm{d}x + b\mathrm{d}y) = \mathrm{d}(\alpha \mathrm{d}x + \beta \mathrm{d}y) = -kH\mathrm{d}x\mathrm{d}y.$$

Therefore, if $\mathrm{d}\omega = 0$, then $k(x,y) = 0$.

Let us prove the equality $\mathrm{d}\omega = 0$. Assume that $\gamma(z,x,y)$ is a geodesic in metric (4.24) and that it connects a point z on the circumference $x^2 + y^2 = 1$, and a point (x,y) in the circle $x^2 + y^2 \le 1$. Let us consider the function

$$w(x,y,z) = \int_{\gamma(z,x,y)} \omega = \int_{\gamma(z,x,y)} a\mathrm{d}x + b\mathrm{d}y \tag{4.30}$$

for $0 \le z \le 2\pi$, $x^2 + y^2 \le 1$. It is convenient to assume that $w(x,y,z)$ is a periodic function of z, i. e. $w(x,y,z+2\pi) = w(x,y,z)$. Differentiating (4.30) along γ we have

$$\begin{gathered} \frac{\partial w}{\partial x}\cos\theta + \frac{\partial w}{\partial y}\sin\theta = a\cos\theta + b\sin\theta, \\ \cos\theta = \frac{\mathrm{d}x}{\sqrt{\mathrm{d}x^2 + \mathrm{d}y^2}}, \quad \sin\theta = \frac{\mathrm{d}y}{\sqrt{\mathrm{d}x^2 + \mathrm{d}y^2}}. \end{gathered} \tag{4.31}$$

Besides, from (4.29) and (4.30) for $z = z_1$ and $x = \cos z_2$, $y = \cos z_2$ we have

$$w(z_1, \cos z_2, \sin z_2) = \int_{\gamma(z_1,z_2)} \omega = 0. \tag{4.32}$$

The following equations are valid:

$$\begin{aligned} 0 \;=\; & \left[\left(\frac{\partial w}{\partial x} - a\right)\sin\theta - \left(\frac{\partial w}{\partial y} - b\right)\cos\theta\right] \frac{\partial}{\partial z}\left[\left(\frac{\partial w}{\partial x} - a\right)\cos\theta + \left(\frac{\partial w}{\partial y} - b\right)\sin\theta\right] \\ & - \left[\left(\frac{\partial w}{\partial x} - a\right)\cos\theta + \left(\frac{\partial w}{\partial y} - b\right)\sin\theta\right] \frac{\partial}{\partial z}\left[\left(\frac{\partial w}{\partial x} - a\right)\sin\theta - \left(\frac{\partial w}{\partial y} - b\right)\cos\theta\right] \\ =\; & \left[\left(\frac{\partial w}{\partial x} - a\right)^2 - \left(\frac{\partial w}{\partial y} - b\right)^2\right]\frac{\partial \theta}{\partial z} + \frac{\partial}{\partial y}\left(\frac{\partial w}{\partial x}\frac{\partial w}{\partial z}\right) - \frac{\partial}{\partial x}\left(\frac{\partial w}{\partial y}\frac{\partial w}{\partial z}\right) - \frac{\partial}{\partial z}\left(a\frac{\partial w}{\partial y}\right) + \frac{\partial}{\partial z}\left(b\frac{\partial w}{\partial x}\right). \end{aligned}$$

Hence using the Stokes formula, the periodicity of $w(x,y,z)$ (with respect to z) and (4.32) we integrate to find

$$\int_0^{2\pi}\int_{x^2+y^2\le 1}\left[\left(\frac{\partial w}{\partial x}-a\right)^2-\left(\frac{\partial w}{\partial y}-b\right)^2\right]\frac{\partial\theta}{\partial z}\mathrm{d}x\mathrm{d}y\mathrm{d}z=0. \tag{4.33}$$

According to the conditions of the theorem, any two points in the circle can be connected by the unique geodesic γ, particularly, the points z and (x,y). Therefore, the function $\theta=\theta(x,y,z)$ is strictly monotone with respect to z for each fixed (x,y), $x^2+y^2<1$. Consequently, (4.33) implies

$$\frac{\partial w}{\partial x}=a,\quad \frac{\partial w}{\partial y}=b,$$

which, in turn, yield $\mathrm{d}(a\mathrm{d}x+b\mathrm{d}y)=\mathrm{d}\omega=0$. The theorem is proved.

4.5 THE SOLVABILITY OF A PROBLEM IN INTEGRAL GEOMETRY BY INTEGRATION ALONG GEODESICS

We consider the question of the solvability of a problem in integral geometry in the case when the curves along which the integration of the unknown function is carried out are geodesics of a fixed analytic metric (see 4.2.5). Here we give the proof. Let

$$\mathrm{d}s^2=B^2(x,y)(\mathrm{d}x^2+\mathrm{d}y^2) \tag{4.34}$$

be an analytic metric, defined on the $(x,y)-$ plane. Henceforth we suppose that $B>0$ and $\partial B/\partial y<0$. We denote by $\tilde\gamma(\xi,\varphi)$ the geodesic of the metric (4.34) starting in the point $(\xi,0)$ at an angle φ and let $\gamma(\xi,\varphi)$ be the part of $\tilde\gamma(\xi,\varphi)$ lying in the half-plane $y\ge 0$.

We consider the following problem: in the domain $-\delta\le\xi\le\delta$, $0\le\varphi\le\infty$ the function

$$w(\xi,\varphi)=\int_{\gamma(\xi,\varphi)}\lambda(x,y)\sqrt{\mathrm{d}x^2+\mathrm{d}y^2}$$

is defined, and one is required to find the function $\lambda(x,y)$ in the domain $y\ge 0$.

Before formulating the result, we give some auxiliary material. We denote by $\eta(\xi,\varphi)$, the point of the line $y=0$ that belongs to the geodesic $\gamma(\xi,\varphi)$ and by $\theta(\xi,\varphi)$ the angle that $\gamma(\xi,\varphi)$ makes with the $x-$ axis at the point $(\eta(\xi,\varphi),0)$. We put

$$a(\xi,\varphi)=B(\xi,0)\sin\varphi,\quad b(\xi,\varphi)=B(\eta(\xi,\varphi),0)\sin\theta.$$

The equations of the geodesics of (4.34) for the choice of parameter

$$t=\int_\gamma\frac{\sqrt{\mathrm{d}x^2+\mathrm{d}y^2}}{B}$$

are as follows

$$\frac{\mathrm{d}^2x}{\mathrm{d}t^2}=B\frac{\partial B}{\partial x},\quad \frac{\mathrm{d}^2y}{\mathrm{d}t^2}=B\frac{\partial B}{\partial y},\quad \left(\frac{\mathrm{d}x}{\mathrm{d}t}\right)^2+\left(\frac{\mathrm{d}y}{\mathrm{d}t}\right)^2=\lambda^2. \tag{4.35}$$

Using this notation, we see easily that any first analytic integral of (4.35) can be written in the form

$$\tilde{r}(x,y,x',y') = u(x,y,y')x' + v(x,y,y'),$$

where $u(x,y,y')$ is an even analytic function of y' and $v(x,y,y')$ is an odd analytic function of y'.

Henceforth we consider real analytic functions in a neighbourhood of the origin of a real Euclidean space E^n, $n \geq 1$.

Let F be the set of all analytic functions $f(x,z), x, z \in E^1$. We denote by R the set of analytic functions $r(x,z)$ generated by the first integrals $\tilde{r}$ of (4.35); namely, if

$$\tilde{r} = u(x,y,y')x' + v(x,y,y'), \quad \frac{\mathrm{d}\tilde{r}}{\mathrm{d}t} = 0,$$

is a first integral of (4.35), then $r(x,z) = u(x,0,z) + v(x,0,z)$. Let F/R be the quotient space of the equivalence classes: $f \sim g$ if $f - g \in R$. We define the set A of functions $w(\xi,\varphi)$ by putting for $f \in F$

$$\begin{aligned} w(\xi,\varphi) &= \frac{1}{2}\left[f(\eta,b)(1+B(\eta,0)\cos\theta - f(\eta,-b)(1-B(\eta,0)\cos\theta\right] \\ &\quad -\frac{1}{2}\left[f(\xi,a)(1+B(\xi,0)\cos\varphi - f(\xi,-a)(1-B(\xi,0)\cos\varphi\right]. \end{aligned}$$

Here $\eta(\xi,\varphi)$, $b(\xi,\varphi)$, $a(\xi,\varphi)$ and $\theta(\xi,\varphi)$ are the functions defined above. We may denote a representation of the function $w(\xi,\varphi)$ by Af and write $w = Af$, $f \in F$. We denote by Df, $f(x,z) \in F$ the analytic function $\lambda(x,y)$ of the variables (x,y) defined as follows: we consider the system of equations

$$\begin{aligned} \frac{\partial u_k}{\partial y} &= -\frac{\partial v_k}{\partial x} - 2(k+1)B\frac{\partial B}{\partial y}u_{k+1}, \\ \frac{\partial v_k}{\partial y} &= \frac{\partial u_k}{\partial x} - u_{k+1}B\frac{\partial B}{\partial x} - \frac{\partial u_{k+1}}{\partial x}B^2 - (2k+3)v_{k+1}B\frac{\partial B}{\partial y}, \end{aligned} \tag{$*$}$$

with Cauchy data

$$u_k\big|_{y=0} = u_k^0(x), \quad v_k\big|_{y=0} = v_k^0(x),$$

such that

$$\sum_{k=0}^{\infty}\left(u_k^0(x)z^{2k} + v_k^0(x)z^{2k+1}\right) = f(x,z), \quad f(x,z) \in F;$$

if this problem has an analytic solution $u_k(x,y)$, $v_k(x,y)$, $k = 0,1,2,\ldots$, then by definition

$$Df = u_0(x,y)\frac{\partial B}{\partial x} + v_0(x,y)\frac{\partial B}{\partial y} + \frac{\partial u_0}{\partial x}B.$$

Theorem 4.24.

1. *A necessary and sufficient condition for the solvability of the integral geometry problem in the class of analytic functions $\lambda(x,y)$ is that the function*

$$w(\xi,\varphi) = \int_{\gamma(\xi,\varphi)} \lambda(x,y)\sqrt{\mathrm{d}x^2 + \mathrm{d}y^2}$$

can be represented in the form $w = Af$, $f(x,z) \in F$. For any function $f(x,z) \in F$ the operation Df is defined, and

$$Af = \int_{\gamma(\xi,\varphi)} Df\sqrt{dx^2 + dy^2}.$$

2. *$Af_1 = Af_2$ if and only if the equivalence classes of F/R corresponding to the functions $f_1(x,z)$ and $f_2(x,z)$ coincide.*

3. *If*

$$\int_{\gamma(\xi,\varphi)} \lambda_1\sqrt{dx^2 + dy^2} = \int_{\gamma(\xi,\varphi)} \lambda_2\sqrt{dx^2 + dy^2}$$

and $\lambda_i(x,y)$, $i = 1,2$, are analytic functions, then $\lambda_1(x,y) = \lambda_2(x,y)$.

PROOF. Let $\lambda(x,y)$ be an analytic solution of the integral geomtry problem. We consider the Cauchy problem

$$\frac{\partial \tilde{u}}{\partial z} = -\frac{1}{B\frac{\partial B}{\partial y}}\left[\frac{\partial \tilde{u}}{\partial y}z + \frac{\partial \tilde{v}}{\partial x}\right], \quad \frac{\partial \tilde{v}}{\partial z} = -\frac{1}{B\frac{\partial B}{\partial y}}\left[\frac{\partial \tilde{v}}{\partial y}z + \frac{\partial \tilde{u}}{\partial x}\left(B^2 - z\right)^2 + \tilde{u}B\frac{\partial B}{\partial x} + \lambda B\right], \tag{4.36}$$

$$\tilde{u}\big|_{z=0} = \tilde{u}_0(x,y), \quad \tilde{v}\big|_{z=0} = \tilde{v}_0(x,y), \tag{4.37}$$

where $\tilde{u}_0(x,y)$ and $\tilde{v}_0(x,y)$ are certain analytic functions. Since by hypothesis $B > 0$ and $\partial B/\partial y < 0$, it follows from the Cauchy-Kowalewski theorem that there is a unique analytic solution $\tilde{u}(x,y,z)$, $\tilde{v}(x,y,z)$ of this problem. We put

$$u(x,y,z) = \frac{\tilde{u}(x,y,z) + \tilde{u}(x,y,-z)}{2}, \quad v(x,y,z) = \frac{\tilde{v}(x,y,z) - \tilde{v}(x,y,-z)}{2}. \tag{4.38}$$

It is easy to see that these functions are the solution of (4.36). Since $u(x,y,z)$ and $v(x,y,z)$ are even and odd in z, we can expand them in the following series:

$$u = \sum_{k=0}^{\infty} u_k(x,y)z^{2k}, \quad v = \sum_{k=0}^{\infty} v_k(x,y)z^{2k+1}. \tag{4.39}$$

Substituting these series in (4.36) and comparing coefficients, we obtain

$$\begin{gathered}\frac{\partial u_k}{\partial y} = -\frac{\partial v_k}{\partial x} - 2(k+1)B\frac{\partial B}{\partial y}u_{k+1}, \\ \frac{\partial v_k}{\partial y} = \frac{\partial u_k}{\partial x} - u_{k+1}B\frac{\partial B}{\partial x} - \frac{\partial u_{k+1}}{\partial x}B^2 - (2k+3)v_{k+1}B\frac{\partial B}{\partial y},\end{gathered} \tag{4.40}$$

$$\lambda(x,y) = u_0(x,y)\frac{\partial B}{\partial x} + v_0(x,y)\frac{\partial B}{\partial y} + \frac{\partial u_0}{\partial x}B(x,y). \tag{4.41}$$

We consider the function $g(x,y,x',y')$ defined as follows:

$$g = u(x,y,y')x' + v(x,y,y'),$$

where $u(x,y,z)$ and $v(x,y,z)$ are defined by (4.38). Let the parameter t on the geodesic γ of the metric $\mathrm{d}s^2 = B^2(x,y)(\mathrm{d}x^2+\mathrm{d}y^2)$ be chosen as

$$t = \int_\gamma \frac{\sqrt{\mathrm{d}x^2+\mathrm{d}y^2}}{B}.$$

As we noted above, for this choice of t the system of equations that defines the geodesic γ has the form

$$\frac{\mathrm{d}^2x}{\mathrm{d}t^2} = B\frac{\partial B}{\partial x}, \quad \frac{\mathrm{d}^2y}{\mathrm{d}t^2} = B\frac{\partial B}{\partial y}, \quad \frac{\mathrm{d}x}{\mathrm{d}t} = B\cos\varphi, \quad \frac{\mathrm{d}y}{\mathrm{d}t} = B\sin\varphi, \tag{4.42}$$

where $(\cos\varphi, \sin\varphi)$ is the unit tangent vector of γ. We show that whatever the geodesic $\gamma = \{x(t), y(t)\}$, we have

$$\frac{\mathrm{d}}{\mathrm{d}t}\left[g\left(x, y, \frac{\mathrm{d}x}{\mathrm{d}t}, \frac{\mathrm{d}y}{\mathrm{d}t}\right)\right] = \lambda\left(x(t), y(t)\right) B\left(x(t), y(t)\right).$$

In fact,

$$\frac{\mathrm{d}}{\mathrm{d}t}g = \frac{\partial g}{\partial x}x' + \frac{\partial g}{\partial y}y' + \frac{\partial g}{\partial x'}x'' + \frac{\partial g}{\partial y'}y''.$$

Since $g = ux' + v$ and $x'' = B\dfrac{\partial B}{\partial x}$, $y'' = B\dfrac{\partial B}{\partial y}$ it follows that

$$\frac{\mathrm{d}g}{\mathrm{d}t} = \left(\frac{\partial u}{\partial x}x' + \frac{\partial u}{\partial x}\right)x' + \left(\frac{\partial u}{\partial y}x' + \frac{\partial u}{\partial y}\right)y' + uB\frac{\partial B}{\partial x} + \left(\frac{\partial u}{\partial y'}x' + \frac{\partial u}{\partial y'}\right)B\frac{\partial B}{\partial y}.$$

Replacing x'^2 by $B^2 - y'^2$ and using the expansions (4.39) of $u(x,y,y')$ and $v(x,y,y')$ as series in y', we have

$$\frac{\mathrm{d}g}{\mathrm{d}t} = x'\left(\sum_{k=0}^{\infty}\left[\frac{\partial u_k}{\partial y} + \frac{\partial v_k}{\partial x} + 2(k+1)u_{k+1}B\frac{\partial B}{\partial y}\right]y'^{2k+1}\right)$$

$$+\sum_{k=0}^{\infty} y'^{2k+2}\left(\frac{\partial v_k}{\partial y} - \frac{\partial u_k}{\partial x} + u_{k+1}B\frac{\partial B}{\partial x} + (2k+3)v_{k+1}B\frac{\partial B}{\partial y} + \frac{\partial u_{k+1}}{\partial x}B^2\right)$$

$$+u_0B\frac{\partial B}{\partial x} + v_0B\frac{\partial B}{\partial y} + \frac{\partial u_0}{\partial x}B^2.$$

The functions $u_k(x,y)$ and $v_k(x,y)$ satisfy the system of equations (4.40), and so

$$\frac{\mathrm{d}g}{\mathrm{d}t} = u_0B\frac{\partial B}{\partial x} + v_0B\frac{\partial B}{\partial y} + \frac{\partial u_0}{\partial x}B^2. \tag{4.43}$$

By what we proved above (see (4.41)),

$$u_0B\frac{\partial B}{\partial x} + v_0B\frac{\partial B}{\partial y} + \frac{\partial u_0}{\partial x}B^2 = \lambda(x,y)B.$$

Consequently (4.43) can be written in the form

$$\frac{dg}{dt} = \lambda B.$$

Let $f(x,z) = u(x,0,z) + v(x,0,z)$. We then have

$$u(x,0,z) = \frac{f(x,z)+f(x,-z)}{2}, \quad v(x,0,z) = \frac{f(x,z)-f(x,-z)}{2}. \tag{4.44}$$

Integrating $\mathrm{d}g = \lambda B \mathrm{d}t$ along the geodesic $\gamma(\xi,\varphi)$, we obtain

$$-g(x,0,x',y')\big|_{x=\eta(\xi,\varphi)}^{x=\xi} = \int\limits_{\gamma(\xi,\varphi)} B\mathrm{d}t,$$

the left-hand side of which is equal to

$$g(\eta,0,B(\eta,0)\cos\theta, B(\eta,0)\sin\theta) - g(\xi,0,B(\xi,0)\cos\varphi, B(\xi,0)\sin\varphi).$$

Substituting for $g|_{y=0}$ its expression $g|_{y=0} = (ux'+v)|_{y=0}$ and using (4.42), we obtain

$$-g(x,0,x',y')\big|_{x=\xi}^{x=\eta(\xi,\varphi)} = Af.$$

Since $B\mathrm{d}t = \sqrt{\mathrm{d}x^2+\mathrm{d}y^2}$ along γ and the function $\lambda(x,y)$ is defined by (4.41), we have

$$Af = \int\limits_{\gamma(\xi,\varphi)} Df\sqrt{\mathrm{d}x^2+\mathrm{d}y^2}$$

Now let

$$f(x,z) = \sum_{k=0}^{\infty}\left(u_k^0(x)z^{2k} + v_k^0(x)z^{2k+1}\right)$$

be a fixed analytic function, and $w(\xi,\varphi) = Af$. We need to prove that the Cauchy problem

$$\begin{gathered}\frac{\partial u_k}{\partial y} = -\frac{\partial v_k}{\partial x} - 2(k+1)B\frac{\partial B}{\partial y}u_{k+1},\\ \frac{\partial v_k}{\partial y} = \frac{\partial u_k}{\partial x} - u_{k+1}B\frac{\partial B}{\partial x} - \frac{\partial u_{k+1}}{\partial x}B^2 - (2k+3)v_{k+1}B\frac{\partial B}{\partial y},\end{gathered} \tag{4.45}$$

$$v_k|_{y=0} = v_k^0(x), \quad u_k|_{y=0} = u_k^0(x), \quad k=0,1,2,\dots, \tag{4.46}$$

has a unique analytic solution $u_k(x,y)$, $v_k(x,y)$, where the series

$$\sum_{k=0}^{\infty} u_k z^{2k} + v_k z^{2k+1}$$

converges. When proved, then the analytic function

$$\lambda(x,y) = Df = u_0\frac{\partial B}{\partial x} + v_0\frac{\partial B}{\partial y} + \frac{\partial u_0}{\partial x}B$$

is a solution of the integral geometry problem for $w = Af$. The uniqueness of the analytic solution of the Cauchy problem (4.45), (4.46) is proved by the usual method. The proof of the existence is carried out by the method of successive approximations with the use of majorants. Let $f(x,z)$ be analytic in the domain $|x| \le x_0$, $|z| \le z_0$. We consider the function $\tilde{f}(x,t) = f(x,\varepsilon z)$, $\varepsilon z = t$, where $\varepsilon > 0$ is an arbitrary number. The function $\tilde{f}(x,t)$ is analytic in the domain $|x| \le x_0$, $|t| \le z_0/\varepsilon$ and has majorant

$$f(x,t) \le \frac{M_1}{\left(1-\dfrac{x}{x_0}\right)\left(1-\dfrac{t}{t_0}\right)}, \tag{4.47}$$

where M_1, x_0 and $t_0$are constants. We choose $\varepsilon > 0$ so that $1/t_0 < 1$. We put $1/t_0 = b$. It obviously follows from (4.47) that if $\tilde{f}(x,t) = \sum_0^\infty f_k(x)t^k$, then $f_k(x)$ is majorized by the function

$$\frac{Mb^k}{\left(1 - \frac{x}{x_0}\right)}, \quad k = 0, 1, 2, \ldots, \quad b < 1.$$

We consider the system of equations

$$\begin{aligned} \frac{\partial \tilde{u}_k}{\partial y} &= \frac{\partial \tilde{v}_k}{\partial x} - \frac{2(k+1)}{\varepsilon} B \frac{\partial B}{\partial y} \tilde{u}_{k+1}, \\ \frac{\partial \tilde{v}_k}{\partial y} &= \frac{\partial \tilde{u}_k}{\partial x} - \frac{1}{\varepsilon}\left[\tilde{u}_{k+1} B \frac{\partial B}{\partial x} - \frac{\partial \tilde{u}_{k+1}}{\partial x} B^2 - (2k+3)\tilde{v}_{k+1} B \frac{\partial B}{\partial y}\right] \end{aligned} \tag{4.48}$$

with Cauchy data

$$\tilde{u}_k\big|_{y=0} = \tilde{u}_k^0(x), \quad \tilde{v}_k\big|_{y=0} = \tilde{v}_k^0(x), \tag{4.49}$$

such that

$$\left|\tilde{u}_k^0\right| \le \frac{M_1 b^{2k}}{\left(1 - \frac{x}{x_0}\right)}, \quad \left|\tilde{v}_k^0\right| \le \frac{M_1 b^{2k+1}}{\left(1 - \frac{x}{x_0}\right)}.$$

Suppose that the functions B^2, $B\frac{\partial B}{\partial x}$ and $B\frac{\partial B}{\partial y}$ are majorized by

$$\frac{M_2}{\left(1 - \frac{x+y}{a}\right)},$$

where $a > 0$ is some number and $a \le x_0$. For any α, $0 < \alpha < 1$, the function

$$\frac{M_2}{\left(1 - \frac{x + \frac{y}{\alpha}}{a}\right)}$$

is also the majorant for B^2, $B\frac{\partial B}{\partial x}$ and $B\frac{\partial B}{\partial y}$. Since $b \ge 1$, the solution $\tilde{u}_k(x,y)$, $\tilde{v}_k(x,y)$ of the problem (4.48), (4.49) is obviously majorized by the solution $\varphi_k(x,y)$, $\psi_k(x,y)$ of the following problem:

$$\begin{aligned} \frac{\partial \varphi_k}{\partial y} &= \frac{M_3}{\left(1 - \frac{x + \frac{y}{\alpha}}{a}\right)} \left[\frac{\partial \psi_k}{\partial x} + (k+1)\varphi_{k+1}\right], \\ \frac{\partial \psi_k}{\partial y} &= \frac{M_3}{\left(1 - \frac{x + \frac{y}{\alpha}}{a}\right)} \left[\frac{\partial \varphi_k}{\partial x} + (k+1)\psi_{k+1}\right], \\ \varphi_k\big|_{y=0} &= \frac{M_3 b^{2k}}{\left(1 - \frac{x}{a}\right)}, \quad \psi_k\big|_{y=0} = \frac{M_3 b^{2k+1}}{\left(1 - \frac{x}{a}\right)}. \end{aligned} \tag{4.50}$$

Here M_3 is a constant $M_3 \geq \max\left(\frac{10M_2}{\varepsilon}, M_1\right)$. We look for a solution of the problem (4.50) in the form $\varphi_k(p) = \psi_k(p) = q_k(p)$, where $p = \frac{y}{\alpha} + x$. We have

$$\frac{\mathrm{d}q_k}{\mathrm{d}p} = A(p)(p+1)q_{k+1}, \quad A(p) = \frac{M_3}{\left(1-\frac{p}{a}\right)\left(\frac{1}{\alpha} - \frac{M_3}{\left(1-\frac{p}{a}\right)}\right)}.$$

We chose α, β and $M \geq M_3$ so that in the domain $|p| < \beta < a$ we have

$$\left(\frac{1}{\alpha} - \frac{M_3}{\left(1-\frac{p}{a}\right)}\right) > 0, \quad |A(p)| \leq M.$$

We conside the recurrent system of equations

$$\frac{\mathrm{d}q_k^{j+1}}{\mathrm{d}p} = A(p)(p+1)q_{k+1}^j, \quad j = 0,1,2,\ldots, \quad q_k^0 = \frac{Mb^{2k}}{\left(1-\frac{p}{a}\right)}.$$

We suppose that for some j we have

$$\left|q_k^j\right| \leq \frac{M^{j+1}a^j(k+1)(k+2)\ldots(k+j+1)b^{k+j}}{\left(1-\frac{p}{a}\right)^j j!},$$

and show that this inequality is satisfied for $j+1$ also. For,

$$\left|\frac{\mathrm{d}q_k^{j+1}}{\mathrm{d}p}\right| \leq \frac{M^{j+2}a^j(k+1)(k+2)\ldots(k+j+2)b^{k+1+j}}{\left(1-\frac{p}{a}\right)^j j!}.$$

Integrating this inequality with respect to p, we have

$$\left|q_k^{j+1}\right| \leq \frac{M^{j+2}a^{j+1}(k+1)(k+2)\ldots(k+j+2)b^{k+1+j}}{\left(1-\frac{p}{a}\right)^{j+1}(j+1)!},$$

which proves the assertion. We now show that the series

$$\sum_{k=0}^{\infty}\sum_{j=0}^{\infty}\frac{M^{j+1}a^j(k+1)(k+2)\ldots(k+j+1)b^{k+j}}{\left(1-\frac{p}{a}\right)^j j!} \tag{4.51}$$

converges if a is sufficiently small. Suppose that $\frac{Mab}{\left(1-\frac{p}{a}\right)} = \tilde{b}$, and a are such that $\tilde{b} + b < 1$ for $p \leq a^2$. Such a exists, since $b < 1$. In this notation the series (4.51) can be rewritten in the form

$$M\sum_{k=0}^{\infty}\sum_{j=0}^{\infty}\frac{(k+1+j)!}{k!j!}\tilde{b}^j b^k,$$

but the sum of this series is equal to $\dfrac{M}{\left[1-(b+\tilde{b})\right]^2}$. We thus have shown that the series (4.51) converges, and thereby we have proved the existence of a solution $\tilde{u}_k(x,y)$, $\tilde{v}_k(x,y)$ of the problem (4.48), (4.49).

We put

$$u_k(x,y)=\varepsilon^k\tilde{u}_k(x,y),\quad v_k(x,y)=\varepsilon^k\tilde{v}_k(x,y),$$

where $\tilde{u}_k$, $\tilde{v}_k$ is a solution of the problem (4.48), (4.49). Obviously $u_k(x,y)$ and $v_k(x,y)$ are analytic and are a solution of the problem (4.45), (4.46). This proves the first assertion of the theorem.

We now prove the second assertion of the theorem. Suppose that the equivalence classes of the space F/R corresponding to the functions $f_1(x,z)$ and $f_2(x,z)$ coincide. We show that $Af_1=Af_2$. Since $f_1(x,z)$ and $f_2(x,z)$ determine one element of F/R we have $f_1-f_2=r(x,z)$, where $r(x,z)$ is defined by the first integral $ux'+v$ of the system (4.35). Along $\gamma(\xi,\varphi)$ we have

$$\frac{\mathrm{d}}{\mathrm{d}t}(ux'+v)=0.$$

Integrating this ,we obtain $Ar=0$. Therefore $Ar=A(f_1-f_2)=Af_1-Af_2=0$, that is, $Af_1=Af_2$.

We now suppose that $Af_1=Af_2$ and show that the equivalence classes corresponding to the functions $f_1(x,z)$ and $f_2(x,z)$ coincide. We consider the difference $r(x,z)=f_1-f_2$. The function $r(x,z)$, as the difference of two analytic functions, is analytic. From this and what was proved above it follows that there is a solution $u_k(x,y), v_k(x,y)$ of the system with Cauchy data

$$u_k^0(x)=r_{2k}(x),\quad v_k^0(x)=r_{2k+1}(x),\quad r=\sum_{k=0}^{\infty}r_k(x)z^k \tag{4.52}$$

and we have

$$Af=\int\limits_{\gamma(\xi,\varphi)}Dr\sqrt{\mathrm{d}x^2+\mathrm{d}y^2}.$$

Since $Ar=Af_1-Af_2=0$ whatever the geodesic $\gamma(\xi,\varphi)$, $|\xi|<\delta$, $0\le\varphi\le\infty$ we have

$$\int\limits_{\gamma(\xi,\varphi)}Dr\sqrt{\mathrm{d}x^2+\mathrm{d}y^2}=0. \tag{4.53}$$

The function Dr is analytic, so it follows from (4.53) that $Dr=0$. But this in turn implies that

$$\frac{\mathrm{d}}{\mathrm{d}t}(ux'+v)=0$$

along the geodesic, where

$$u=\sum_{k=0}^{\infty}u_k(x){y'}^{2k},\quad v=\sum_{k=0}^{\infty}v_k(x){y'}^{2k+1},$$

and $u_k(x,y), v_k(x,y)$ is the solution of the system (4.45) with data (4.52). In other words, the function $ux'+v$ is a first integral of the system (4.35). Consequently $r(x,z)\in R$. This proves the second assertion.

Since the function $\lambda(x,y) = \lambda_1(x,y) - \lambda_2(x,y)$ is analytic, the third assertion of the theorem follows from Theorem 1.1.

REMARK . In the case $f(x,z) = u_0(x) + v_0(x)z$, $x < \delta$ the system of equations (*) is transformed into the system of Cauchy-Riemann equations

$$\frac{\partial u}{\partial y} = -\frac{\partial v}{\partial x'}, \quad \frac{\partial v}{\partial y} = \frac{\partial u}{\partial x},$$

and we have

$$\lambda(x,y) = Df = u\frac{\partial B}{\partial x} + v\frac{\partial B}{\partial y} + \frac{\partial u}{\partial x}B.$$

In particular, it follows that the solution of the integral geometry problem considered above is unstable. On the one hand, small changes in $u_0(x)$ and $v_0(x)$ cause small changes in $w = Af$, since w is defined by $u_0(x)$ and $v_0(x)$ alone. On the other hand, in order to find a solution $\lambda = Df$, we need to carry out an analytic continuation, and the solution of this problem is unstable.

References

Ambarzumjan W.A. (1929). Über eine Frage der Eigenwerttheorie. *Z. für Physik* **53**, 690–695.

Aleksandrov A.D. (1937a). To the theory of mixed volumes. *Mat. Sb. Acad. Nauk SSSR* **44**, 947–972.

Aleksandrov A.D. (1937b). To the theory of mixed volumes. *Mat. Sb. Acad. Nauk SSSR* **44**, 1205–1238.

Aleksandrov A.D. (1938a). To the theory of mixed volumes. *Mat. Sb. Acad. Nauk SSSR* **45**, 27–44.

Aleksandrov A.D. (1938b). To the theory of mixed volumes. *Mat. Sb. Acad. Nauk SSSR* **45**, 228–251.

Alekseev B.V. (1983). *Mathemetical Kinetics of Reacting Gases.* Nauka, Moscow (in Russian).

Anger G. (1990). *Inverse Problems in Differential Equations. Mathematische Lehrbuchen und Monographien. Mathematische Monographien*, 79. Akademie-Verlag, Berlin.

Anikonov D.S. (1975). On the uniqueness in determining the coeffient in the right side of transport equation. *Sib. Math. J.* **16**, 329–334.

Anikonov D.S., Prohorov I.V., and Kovtanjuk A.E. (1993). Diagnostics of scattering and absorbing medium by methods of X-ray tomography. *J. of Inv. and Ill-posed Probl.* **1**, 259–281.

Anikonov Yu.E. (1968). Remarks on convex surfaces. *Sib. Math. J.* **9**, 1413–1415 (in Russian).

Anikonov Yu.E. (1969a). *On Certain Problems of Calculus of Variations and Integral Geometry.* Preprint, Computer Center, SO AN SSSR, Novosibirsk (in Russian).

Anikonov Yu.E. (1969b). Inversion formulae for problems of kinematic seismic and integral geometry. In: *Mathematical Problems of Geophysics*, N 1, Computer Center, SO AN SSSR, Novosibirsk, pp.41–48 (in Russian).

Anikonov Yu.E. (1969c). A uniqueness theorem for convex surfaces. *Math. Notes* **6**, 528–529.

Anikonov Yu.E. (1978a). On solvability of problem in integral geometry. *Math. USSR Sb.* **30**, 239–247.

Anikonov Yu.E. (1978b). *Some Methods of Studying Multidimensional Inverse Problems for Differential Equations.* Nauka, Novosibirsk (in Russian).

Anikonov Yu.E. (1982). On problems of integral geometry. In: *Questions of Correctness of Inverse Problems in Mathematical Problems.* Computer Center, SO AN SSSR, Novosibirsk, pp. 20–29 (in Russian).

Anikonov Yu.E. (1983). On uniqueness classes of solution for problems of integral geometry. In: *Methods of Investigation of Incorrect Problems of Mathematical Physics.* Computer Center, SO AN SSSR, Novosibirsk, pp. 16–17 (in Russian).

Anikonov Yu.E. (1985). Some problems of the theory of inverse problems for kinetic equations. In: *Inverse Problems of Mathematical Physics.* Lavrent'ev M.M., Ed., Computer Center, SO AN SSSR, Novosibirsk, pp. 28–41 (in Russian).

Anikonov Yu.E. (1986a). *Pseudodifferential Operators and Inverse Problems.* Preprint N.86–671. Computer Center, SO AN SSSR, Novosibirsk (in Russian).

Anikonov Yu.E. (1986b). An inverse problem for kinetic equation. In: *Uniqueness and Stability of Some Problems of Mathematical Physics.* Lavrent'ev M.M., Ed., Computer Center, SO AN SSSR, Novosibirsk, pp. 16–22 (in Russian).

Anikonov Yu.E. (1987). New mathematical models and inverse problems. In: *Mathematical Modelling in Geophysics.* Alekseev A.S., Ed., Computer Center, SO AN SSSR, Novosibirsk, pp. 3–42 (in Russian).

Anikonov Yu.E. (1989). A linear system of equations of transport theory which does not contain sections or a scattering indicatrix. *Sov. Phys. Dokl.* **34**, 210–211.

Anikonov Yu.E. (1991). On uniqueness of the solution of an inverse problem for a quantum kinetic equation. *Math. USSR Sb.* **69**, 69–76.

Anikonov Yu.E. and Amirov A.Kh. (1983). A uniqueness theorem for the solution of an inverse problem for a kinetic equation. *Sov. Math. Dokl.* **28**, 510–511.

Anikonov Yu.E. and Belov Yu.Ya. (1989). On unique solvability of an inverse problem for a parabolic equation. *Sov. Math. Dokl.* **39**, 601–605.

Anikonov Yu.E. and Bubnov B.A. (1988). Inverse problems of transport theory. *Sov. Math. Dokl.* **37**, 497–499.

Anikonov Yu.E. and Bubnov B.A. (1989). Control and inverse problems. *Sov. Phys. Dokl.* **34**, 13–14.

Anikonov Yu.E. and Golubyatnikov V.P. (1978). To the question of the uniqueness of a solution of inverse problems of the scattering theory. In: *Inverse Problems for Differential Equations of Mathematical Physics.* Computer Center, SO AN SSSR, Novosibirsk, pp. 13–17 (in Russian).

Anikonov Yu.E. and Marchuk A.G. (1975). To the inverse problem of diffraction. In: *Mathematical Problems of Geophysics,.* N 6. Computer Center, SO AN SSSR, Novosibirsk, pp. 54–62 (in Russian).

Anikonov Yu.E. and Pestov L.N. (1990a). Integral geometry and the structure of Riemannian spaces. *Sov. Math. Dokl.* **40**, 90–93.

Anikonov Yu.E. and Pestov L.N. (1990b). *Formulas in Linear and Nonlinear Problems of Tomography.* Novosibirsk State University, Novosibirsk (in Russian).

Anikonov Yu.E. and Romanov V.G. (1983). On uniqueness form of degree one from its integral along geodesics. In: *Incorrect Mathematical Problems and Problems of Geophysics.* Computer Center, SO AN SSSR, Novosibirsk, pp.22–27 (in Russian).

Anikonov Yu.E. and Shneiberg I.Ya. (1991). Radon transformation with variable absorption, *Dokl. Akad. Nauk SSSR* **318**, 265–267 (in Russian).

Anikonov Yu.E. and Stepanov V.N. (1982). Inversion formula of one integral equation of the first kind and its applications. In: *Methods of Solution of Ill-posed Problems and its Applications.* Computer Center, SO AN SSSR, Novosibirsk, pp. 171–174 (in Russian).

Anikonov Yu.E. and Stepanov V.N. (1983). Uniqueness and stability of the solution of a problem of geometry in the large. *Math. USSR Sb.* **44**, 483–490.

Anikonov Yu.E. and Stepanov V.N. (1991). Inversion formula in one problems of integral geometry. *Dokl. Akad. Nauk SSSR* **316**, 93–95 (in Russian).

Bakel'man I.Ya. (1965). *Geometrical Methods of Solution of Elliptic Equations.* Nauka, Moscow (in Russian).

Bakel'man I.Ya., Verner A.L., and Kantor B.E. (1973). *Introdution in Differential Geometry in the Large.* Nauka, Moscow (in Russian).

Belov Yu.Ya. (1986). *The Method of Weak Approximation for Splitting Parabolic Equations.* Preprint. Krasnoyarsk (in Russian).

Berezanskij Yu.M. (1958). The uniqueness theorem in the inverse problem of the spectral analysis for the Schrödinger equation. *Trudy Moskov. Mat. Obshch.* **7**, 1–62 (in Russian).

Blagoveshchenskii A.S. (1986). Reconstruction of a function from known its integral taken along linear manifolds. *Math. Notes* **39**, 457–461.

Blaschke W. (1916). *Kreis und Kugel.* Veit, Leipzig.

Blaschke W. (1930). *Vorlesungen über Differentialgeometrie und geometrische Grundlagen Einsteins Relativitatstheorie. VI. Elementare Differentialgeometrie.* Springer Verlag, Berlin.

Bogolyubov N.N. (1981). *Lectures on Quantum Statistics.* Nauka, Moscow (in Russian).

Borg G. (1945). Eine Umkehrung der Sturm-Liouvillschen Eigenwertaufgabe, Bestimmung der Differentialglichung durch die Eigenwerte. *Acta Math.* **78**, 1–96.

Bubnov B.A. (1984). A mixed problem for a class of nonclassical equations. *Sov. Math. Dokl.* **29**, 238–241.

Bubnov B.A. (1987a). *On the question of solvability of multidimensional inverse problems for hyperbolic equations.* Preprint N.87–713. Computer Center, SO AN SSSR, Novosibirsk (in Russian).

Bubnov B.A. (1987b). *On the question of solvability of multidimensional inverse problems for parabolic equations.* Preprint N.87–714. Computer Center, SO AN SSSR, Novosibirsk (in Russian).

Bukhgeim A.L. (1972). On some problems of integral geometry. *Sib. Math. J.* **13**, 23–29.

Bukhgeim A.L. (1979). On the solvability of a certain problems of integral geometry, In: *Incorrect Mathematical Problems and Problems of Geophysics.* Computer Center, SO AN SSSR, Novosibirsk, pp. 37–46 (in Russian).

Bukhgeim A.L. (1983). *Volterra Equations and Inverse Problems.* Nauka, Novosibirsk (in Russian).

Bukhgeim A.L. (1986). *Difference Methods for Solving Ill-posed Problems.* Computer Center, SO AN SSSR, Novosibirsk (in Russian).

Bukhgeim A.L. (1988). *Introduction to the Theory of Inverse Problems.* Nauka, Novosibirsk (in Russian).

Busemann H. (1958). *Convex surfaces.* Interscience Publishers, New York.

Campi S. (1986). Reconstructing a convex surface from certain measurements of its projections. *Bolletino Union Math. Ital.* **6**, 945–959.

Case K.M. and Zweifel P.F. (1967). *Linear Transport Theory.* Addison Wesley, London.

Courant R. and Hilbert D. (1962). *Methods of mathematical physics, Vol II: Partial differential equations.* Interscience Publishers, New York-London.

Favard J. (1957). *Cours de Géométrie Différentielle Locale.* Gautier-Villars, Paris.

Ferziger J.H. and Kaper H.G. (1972). *Mathematical Theory of Transport Processes in Gases.* North-Holland, Amsterdam.

Fikhtengol'ts G.M. (1966). *Differential- und Integralrechung. Hochschulbücher für Mathematik*, Band 62, VEB Deutscher Verlag Wiss., Berlin.

Finch D.V. (1985). Cone-beam reconstruction with sources on curve. *SIAM J. Appl. Math.* **45**, 665–673.

Funk P. (1916). Über eine geometrische Anwendung der Abelschen Integralgleichung. *Math. Ann.* **77**, 129–135.

Gel'fand I.M. (1960). Integral geometry and its relation to the theory of representations. *Russian Math. Surveys* **15**, 143–151.

Gelfand I.M., Gindikin S.G., and Graev M.I. (1980). Integral geometry in affine and projective spaces, In: *Current problems in mathematics*, 16, Vsesojuz Inst. Nauch i Tehnc. Informacii, Moscow, pp. 53–226 (in Russian).

Gelfand I.M. and Goncharov A.B. (1987). Recovery of a compactly supported function by its integrals over lines intersecting given set of points in space. *Sov. Math. Dokl.* **34**, 373–376.

Gelfand I.M., Graev M.I., and Shapiro Z.Ya. (1969). Differential forms and integral geometry. *Functional Anal. i Prilozhen* **3**, 24–40 (in Russian).

Gel'fand I.M., Graev M.I., and Vilenkin N.Ya. (1966). *Generalized functions, Vol 5: Integral geometry and representations theory.* Academic Press, New York-London.

Godbillon C. (1969). *Géométrie Différentielle et Méchanique Analitique.* Herman, Paris.

Goncharov A.B. (1988). Methods of integral geometry and recovering function with compact support from its projections in unknown directions. *Acta Appl. Math.* **11**, 213–222.

Gradshtein I.S. and Ryzhik I.M. (1980). *Table of integrals, series and products.* Academic Press, New York,London.

Haken G. (1987). *Advanced Synergetics, Ser. Synergetics*, Vol.20, Springer, Berlin.

Helgason S. (1959). Differential operators on homogeneous spaces. *Acta Math.* **102**, 239–299.

Helgason S. (1980). *The Radon Transform.* Birkhaser Verlag, Boston.

Herglotz G. (1905). Über die Elastizität der Erde bei Berücksichtigung ihrer variablen Dichte. *Z. für Mathematik und Physik* **52**, 275–299.

Isakov V. (1990). *Inverse Source Problem. Mathematical Surveys and Monographs.* 34, American Mathematical Society, Providence, RI.

Ivanov V.K. (1963). On ill-posed problems. *Math. Sb. (N.S.)* **61**(103), 211–223 (in Russian).

John F. (1955). *Plane Waves and Spherical Means Applied to Partial Differential Equations.* Interscience, New York.

Keifitz N. (1968). *Introduction to the Mathematics of Population.* Addison-Wesley, New York.

Khachaturov A.A. (1954). Determination of the value of the measure for a region of $n-$ dimensional Euclidean space from its value for all half-spaces. *Uspehi Matem. Nauk (N.S.)* **9**, 205–212 (in Russian).

Kirillov A.A. (1961). On a problem of I.M.Gelfand. *Sov. Math. Dokl.* **2**, 268–269.

Klimontivich Yu.D. (1982). *Statistical Physics.* Nauka, Moscow. (in Russian).

Kolmogorov A.N. (1985). The qualitative study of mathematical models of the dynamics of populations. In: *Kolmogorov A.N. Mathematics and Mechanics, Selected Works.* Nikol'skii S.M., Ed., Nauka, Moscow, pp. 357–362 (in Russian).

Kolmogorov A.N., Petrovskii I.G., and Piskunov N.S. (1985). The analisis of diffusion equations, combined with the growth of the quantity of material, and its application to a biological problem. In: *Kolmogorov A.N. Mathematics and Mechanics, Selected Works.* Nikol'skii S.M. Ed., Nauka, Moscow, pp. 221–246 (in Russian).

Kompaneets A.S. (1977). *Physico-Chemical and Relativistic Gas Dynamics.* Nauka, Moscow (in Russian).

Kostelyanec P.O. and Reshetnyak Yu.G. (1954). The determination of a completely additive function by its values on half-spaces. *Uspehi Matem. Nauk (N.S.)* **9**, 135–140 (in Russian).

Kuchment P. (1989). *Paley-Wiener theory for exponential Radon transform on the plane* Lecture on the Joint Summer Research conference on Integral Geometry, Arcafa, USA.

Lavrent'ev M.M. (1955). On Cauchy's problem for Laplace equation. *Dokl. Acad. Nauk SSSR (N.S.)* **102**, 205–206 (in Russian).

Lavrent'ev M.M. (1959). Integral equation of the first kind. *Dokl. Acad. Nauk SSSR* **127**, 31–33 (in Russian).

Lavrent'ev M.M. and Anikonov Yu.E. (1967). A certain class of problems in integral geometry. *Dokl. Akad. Nauk SSSR* **176**, 1002–1003 (in Russian).

Lavrent'ev M.M., Anikonov Yu.E., and Fazylov F.N. (1971). Approximate solution of some nonlinear operator equations. *Sov. Math. Dokl.* **12**, 1473–1476.

Lavrent'ev M.M. and Bukhgeim A.L. (1973). On a class of problems of integral geometry. *Sov. Math. Dokl.* **14**, 750–760.

Lavrent'ev M.M. and Romanov V.G. (1966). Three linearized inverse problems for hyperbolic equation. *Sov. Math. Dokl.* **7**, 1650–1652.

Lavrent'ev M.M., Romanov V.G., and Shishatskii S.P. (1986). *Ill-Posed Problems in Mathematical Physics and Analysis Translations of Mathematical Monographs*, 64, American Mathematical Society, Providence RI.

Lavrent'ev M.M., Romanov V.G., and Vasiliev V.G. (1970). *Multi-dimensional Inverse Problems for Differential Equations, Lecture Notes in Mathematics*, 167, Springer-Verlag, Berlin-New York.

Leahy J.V., Smith K.I., and Solmon D.C. (1979). Uniqueness, Nonuniqueness and Inversion in the X-ray and Radon Problems. *Proc. Inter. Symp. on Ill-Posed Problems*, Newark, D.E.

Levis R.M. (1969). Physical optics inverse diffraction. *IEEE Trans. Anten. and Propag.* **17**, 308–314.

Lifshitz and Pitaevsij L.P. (1979). *Physical Kinetics*. Nauka, Moscow (in Russian).

Logvinenko V.N. (1988). Multi-dimensional analysis theorem of Kaztrait and Planjerely-Polia and their applications. In: *Complex Analysis*, Krasnoyarsk, pp. 90–98 (in Russian).

Lorenzi A. (1992). An inverse spectral problem for a nonlinear ordinary differential equation. *Applicable Anal.* **46**, 129–143.

Majda A. (1976a). High frequency asymptotic for the scattering matrix and the inverse problem of acoustical scattering. *Comm. Pure and Appl. Math.* **29**, 261–269.

Majda A. (1976b). A representation formula for the scatteing operator and the inverse problem for arbitraay bodies. *Comm. Pure and Appl. Math.* **30**, 164–185.

Majda A. and Tayler M. (1977a). The asymptotic behavior of the diffraction peak in classical scattering, *Comm. Pure and Appl. Math.* **30**, 639–669.

Majda A. and Tayler M. (1977b). Inverse scattering problem for transparent obstacles electromagnetic waves and hyperbolic system. *Comm. in Part. Diff. Eq.* **2**, 395–438.

Malgrange B. (1966). *Ideals of Differentiable Functions*. Oxford Univ. Press, London.

Matheron G. (1975). *Random Sets and Integral Geometry*. Wiley, New York.

Matsumura S. (1926). Eine charakteristische Eigenschaft der Kugel über Dentsch. *Math. Verlinig* **35**, 298–300.

Mukhometov R.G. (1977). The problem of reconstructing a two-dimensional Riemannian method and integral geometry. *Sov. Math. Dokl.* **18**, 27–31.

Murray J.D. (1977). *Lectures on Nonlinear-Differential Equation Models in Biology*. Clarendon, Oxford.

Natterer F. (1986). *The Mathematics of Computerized Tomography*. Teubner and John Wiley, Stuttgart.

Novikov P.S. (1938). On the uniqueness of the inverse problem of potential theory. *Dokl. Acad. Nauk SSSR* **18**, 1016–1019 (in Russian).

Palamodov V.P. and Denisjuk A.S. (1988). Inversion de la transformation de Radon d'après des données incomplétes. *C.R.Acad. Sci. Paris* **307**, 181–183.

Pestov L.N. (1985). The problem of integral geometry for a symmetric $m-$ form and its applications in kinematics In: *Inverse Problems of Mathematical Physics.* Computer Center, SO AN SSSR, Novosibirsk, pp. 105–119 (in Russian).

Pestov L.N. (1986). On the uniqueness in determining the right side of transport equation. In: *Questions of Well-Posedness for Problems of Mathemetical Physics and Analysis.* Lavrent'ev M.M., Ed., Computer Center, SO AN SSSR, Novosibirsk (in Russian).

Pestov L.N. and Sharafutdinov V.A. (1987). Integral geometry of tensor field on manifold of negative curvature. *Sov. Math. Dokl.* **295**, 305–307.

Pestov L.N. and Sharafutdinov V.A. (1988). Integral geometry of tensor field on manifold of negative curvature. *Sib. Math. J.* **29**, 427–441.

Pilant M. and Rundell W. (1986). An inverse problem for a nonlinear-parabolic equation. *Comm. Part. Diff. Eq.* **11**, 445–457.

Pogorelov A.V. (1973). *Extrinsic Geometry of Convex Surfaces. Transl. Math. Monographs*, 35, American Mathematical Society, Providence, RI.

Pogorelov A.V. (1978). *The Minkovski Multidimensional Problem.* V.H.Winston and John Wiley, Washington, D.C.

Pogorelov A.V. (1979). *Hilbert's fourth problem.* Wiley, Washington.

Prigogine I. (1980). *From Being to Becoming: Time and Complexity in the Physical Sciences.* W.H.Freeman, San Francisco.

Prilepko A.I. and Kostin A.B. (1993). On inverse problems of determining a coefficient in a parabolic equation II. *Sib. Math. J.* **34**, 923–937.

Radon J. (1917). Über die Bestimmung von Funktionen durch ihre Integral wärte längs Gewisser Mannigfaltigkeiten. *Ber. Verh. Sachs. Akad. Wiss. Leipzig., Math.-Nat. Kl.* **68**, 262–277.

Romanov V.G. (1967). Function reconstructing by means of integral along family of curves. *Sib. Math. J.* **8**, 1206–1208.

Romanov V.G. (1978). Integral geometry on geodesics of isotropic Riemannian metric. *Sov. Math. Dokl.* **19**, 847–851.

Romanov V.G. (1987). *Inverse Problem of Mathematical Physics.* VNU Science Press, Utrecht.

Ronkin L.I. (1974). *Introduction to the theory of entire functions of several variables. Translations of Mathematical Monographs*, Vol 44, American Mathematical Society, Providence, RI.

Rundell W. (1989). Determining the birth function population. *Math. Pop. Stud.* **4**.

Sabatier P.C. (1990a). Questions of multidimensional nonlinear equations with exponentially confined solutions. *Inv. Probl.* **6**, L29–L32.

Sabatier P.C. (1990b). Multidimensional nonlinear Schrödinger equations with exponentially confined solutions. *Inv. Probl.* **6**, L47–L53.

Semyanistyi V.I. (1960). On some integral transformations in Euclidean space. *Dokl. Akad. Nauk SSSR* **134**, 536–539 (in Russian).

Semyanistyi V.I. (1961). Homogeneous functions and some problems of integral geometry in space of constant curvature. *Sov. Math. Dokl.* **2**, 59–62.

Semyanistyi V.I. (1966). Certain problems of integral geometry in pseudo-Euclidean and non-Euclidean spaces. *Trudy Sem. Vektor. Tensor. Anal.* **13**, 244–302 (in Russian).

Sharafutdinov V.A. (1984). *On Symmetric Tensor Fields on a Riemannian Manifold.* Preprint N.539. Computer Center, SO AN SSSR, Novosibirsk (in Russian).

Sharafutdinov V.A. (1989). Radial transformation of symmetric tensor fields. *Sov. Math. Dokl.* **39**, 82–85.

Sobolev S.L. (1963). *Applications of Functional Analysis in Mathematical Physics. Translations of Mathematical Monographs*, Vol 7, American Mathematical Society, Providence, RI.

Sobolev S.L. (1974). *Introduction to the theory of cubature formulae.* Nauka, Moscow (in Russian).

Solman D.C. (1976). X-ray transformation. *J. Math. Anal. Appl.* **56**, 61–83.

Stepanov V.N. (1981). Some questions of the geometry of convex surfaces. In: *Investigation of Correctness of Ill-posed Problems and Some Operator Equations.* Computer Center, SO AN SSSR, Novosibirsk, pp. 65–75 (in Russian).

Stepanov V.N. (1984). Geometric aspects of the inverse scattering problem. *Sib. Math. J.* **25**, 111-121.

Svirizhev Yu.M. and Logofet D.O.(1978). *Stability of Biological Communities.* Nauka, Moscow (in Russian).

Taylor M. (1981). *Pseudodifferential Operators.* Princeton University Press, Princeton.

Tikhonov A.N. (1949). On the uniqueness of the solution of the problem of electric prospecting. *Dokl. Akad. Nauk SSSR (N.S.)* **69**, 797–800 (in Russian).

Tikhonov A.N. (1943). On the stability of inverse problems. *C.R.Acad. Sci. USSR, N. Ser.* **39**, 176–179.

Tikhonov A.N. (1963). On the regularization of ill-posed problems. *Dokl. Acad. Nauk SSSR* **153**, 49–52 (in Russian).

Tikhonov A.N. and Arsenin V.Y. (1977). *Solutions of Ill-posed Problems. Scripta series in Mathematics*, John F., Ed., Willey & Sons, New York.

Tikhonov A.N., Arsenin V.Ya., and Timonov A.A. (1987). *Mathematical Problems in Computerized Tomography.* Nauka, Moscow (in Russian).

Tuy H.K. (1983). An inverse formula for cone-beam reconstruction. *SIAM J. Appl. Math.* **43**, 546–552.

Volkov Yu.A. (1963). *Stability of the solution of the Minkowski problem.* Vestnik LGU, 33-43.

Volterra V. (1931). *Leçons sur les Théorie de la Lutte pour la Vie.* Gauthier- Villar, Paris.

Yamamoto M. (1990). Inverse eigenvalue problem for a vibration of a string with viscous drag. *J. Math. Anal. Appl.* **152**, 20–34.

Yamamoto M. (1992). *Generic Uniqueness and Stability in Some Inverse Parabolic Problems.* Preprint. Dept. of Math., University of Tokyo, Tokyo.

Yanenko N.N. (1971). *The Method of Fractional Steps.* Springer Verlag, Berlin.

Bei Fragen zur Produktsicherheit wenden Sie sich bitte an:
If you have any questions regarding product safety,
please contact:

Walter de Gruyter GmbH
Genthiner Straße 13
10785 Berlin
productsafety@degruyterbrill.com